D1703971

Fibre Cement – Technology and Design

Jan R. Krause

Fibre Cement – Technology and Design

Birkhäuser – Publishers for Architecture
Basel · Berlin · Boston

We would like to express our thanks to the
following institution for its support in the
publication of this book:
FH BOCHUM University of Applied Sciences,
School of Architecture,
AMM – Architecture Media Management

Translation: Raymond Peat, Alford

Editors: Caroline Kleist, Brunswick, Elke Stamm, Berlin
Editorial Assistant: Anne-Christin Wolf, Berlin
Graphic Design, layout and cover: Alexander Butz, Brunswick, Anna Schoida, Berlin,
Alexandra Zöller, Berlin
Drawings: Alexander Butz, Sebastian Latz, Martin Meyke, Jan Pingel, Nico Schwarzer,
Markus Willeke, Brunswick
Front cover: Showroom Heidelberg, Design: Astrid Bornheim, Photo: David Franck
Editor for the Publisher: Andreas Müller, Berlin

This book is also available in a German language edition
(ISBN-13: 978-3-7643-7590-4, ISBN-10: 3-7643-7590-6)

Bibliographic information published by Die Deutsche Bibliothek
Die Deutsche Bibliothek lists this publication in the Deutsche Nationalbibliografie;
detailed bibliographic data is available on the Internet at <http://dnb.ddb.de>.

Part of Springer Science+Business Media
Printed on acid-free paper produced from chlorine-free pulp. TFC ∞
Printed in Germany

ISBN-13: 978-3-7643-7591-1
ISBN-10: 3-7643-7591-4

www. birkhauser.ch

9 8 7 6 5 4 3 2 1

Contents

Culture

The material fibre cement has a cultural history and identity of its own. Since its invention in 1900 fibre cement has always had a characteristic building culture. It is a culture of simplicity and experiment, a culture of everyday buildings and the architectural avant-garde. The identity of the material lies in its unmistakable authenticity in appearance and haptic, not forgetting its very universality of use for roof, facade, finishings, furniture and many unusual applications in garden, landscaping and engineering. For more than one hundred years the material has inspired designers and artists – Pablo Picasso, Fernand Leger and Max Ernst even painted on fibre cement. However it is primarily a material for architects. They have made the most of the possibilities of the material to incorporate its individual meaning in their designs. They have chosen fibre cement as an autonomous, form-determining element or used it in combination with other materials for parts of surfaces. Formats from small 20 x 20mm slates up to 3100 x 1500mm large sheets open diverse design possibilities. The characteristic corrugated sheets have become the epitome of the material and the possibilities of form and colour choice open a range of applications that still remains to be fully explored.

Fibre cement is a material that is as simple as it is universal. It has been used by architects for roofs, facades and interiors for more than 100 years. In addition, the material has proved its worth in quite different fields: in design and art, just as much as in a diverse range of everyday applications. The history of the development of fibre cement products is a story of innovation and experiment. In building construction its scope of uses includes roof sheets, facade panels and stair treads. Civil engineering has large diameter water pipes and underground ventilation ducts made from fibre cement. In German towns, sections of fibre cement pipes are used as advertising pillars. Many public spaces such as squares and gardens have been enhanced in appearance by furniture and plant containers made from fibre cement. Among the everyday articles which have been manufactured from this robust material are crazy-golf courses, outdoor bowling alleys, bird baths and a heavy, solid fibre cement barbecue. Today some of the above applications are no longer in production. This may be because other materials have replaced the fibre cement in these products, or there is no longer any call for them. Yet, it is the thin, grey cement material's very versatility, durability and formability that architects, designers and other creative professionals find so interesting to this day. Dietmar Steiner, Head of the Architecture Centre in Vienna, called fibre cement "an essential technical and cultural construction material and an integral part of contemporary architecture". [1]

Day care centre, Berlin-Spandau, 1998, Georg Augustin & Ute Frank

The story of the invention of fibre cement begins with a typical product of 19th century industry. Ludwig Hatschek (1856-1914), the Austrian inventor of the material, came from a family of beer-brewers. He cashed in his inheritance and acquired a factory in Vöcklabruck, Austria, that made asbestos sheets and gaskets. The natural mineral fibre asbestos was proving of interest for many applications during this time of industrialisation, above all because it was fire-proof. This was also an important factor in house-building as many houses in this rural area were roofed with wooden shingles or straw. In 1893 Hatschek founded "the first Austro-Hungarian asbestos factory" and began to develop new products. For seven years he searched for a fire-resistant product for cladding roofs which would be lighter and longer lasting than clay tiles, cheaper and more universal than slate, and more durable than lead. After he had discovered the right binder in Portland cement, a mixture of burnt limestone, water and sand, he was successful in producing a fire-proof, frost-resistant, lightweight sheet for use as durable roof cladding.

Ever since the Romans invented opus caementitium, it has been known that cement-based articles could carry loads in compression but not in tension. The required tensile resistance in Hatschek's thin, strong cement sheets was provided by natural asbestos mineral fibres, which had been known since antiquity for their particularly high resistance to breaking under tension. These fibres exceed the strength of steel wire by a factor of two and are considered to be nature's strongest fibres. The generic term asbestos refers to natural hydrous silicate minerals. The term "asbestos" comes from the Greek and means "indestructible". Asbestos fibres were used at least 3000 years ago as mineral flax to make fire-proof fabrics and were mined in asbestos quarries. Up to the beginning of the 1980s asbestos was incorporated into several thousand applications and products. When new scientific findings proved in 1976 that fine asbestos dust causes health damage, Stephan Schmidheiny, Chairman of Eternit AG, Switzerland, decided that all Swiss-controlled Eternit companies would cease manufacture of asbestos cement. Under the overall charge of the Swiss and German development laboratories and in close cooperation with Eternit in Austria and Belgium, fibre cement was reinvented using other types of fibre.

Opel used car showroom, Frankfurt, 1960, Günter Lange

Scientific tests confirmed in fact that no danger would arise from installed asbestos cement products for roofs and facades using bound fibres (10% asbestos, 90% cement). And into the 1990s asbestos cement products could still be manufactured and used in Europe if dust prevention and extraction were practised. However, by 1979 Eternit AG were already working closely with the German Federal Ministry for Research and Technology to investigate industrial options for asbestos replacement. In 1981 the Asbestos Cement Trade Association (Wirtschaftsverband Asbestzement e.V.) came to a voluntary agreement with the German federal government on behalf of the industry to abandon asbestos cement production. Over 200 different fibres and fibre mixtures were tested before the first successes finally appeared. Extensive testing was required to investigate the long-term behaviour of the new material and to confirm that the fibres were non-hazardous to health and caused no detrimental effects when installed in buildings. Finally, a synthetic organic fibre made from polyvinyl alcohol (PVA) was introduced as the new reinforcing fibre. The same material had been used in a similar form for surgical stitches. In 1981, the first products of the new asbestos-free fibre cement generation were manufactured.

Ludwig Hatschek's invention was less of a product and much more a process. His patent application, dated 28 March 1900, was for "a process for the manufacture of artificial stone sheets with hydraulic binders using fibrous materials". The remarkable fact about this process is that the principle remained unaltered even after the change from natural asbestos fibres to synthetic organic plastic fibres. Fibre cement panels are still produced today using the Hatschek process on Hatschek machines.
Hatschek sold licences for his process all over the world. The name he chose for his material, Eternit, was derived from the Latin "aeternitas" meaning everlasting. By 1910, only a decade after the patent, there were Eternit factories in France, Switzerland, Germany, Austria-Hungary, Belgium, Holland, Portugal, Italy, UK, Sweden, Denmark, Rumania, Russia, USA and Canada. The breakthrough in Austria-Hungary came with the construction of the Tauern mountain railway link from Salzburg to Trieste in 1908. All the station buildings, stores, signal boxes and maintenance facilities along the route were clad in fibre cement sheets. The sparks from the steam locomotives provided a stern test for the new material used for the roof cladding on the track-side buildings.

At the beginning fibre cement was seen as a durable replacement for wooden shingles on roofs, especially on sides exposed to the weather. At the same time, however, architects were also discovering a much sought-after veracity in the material. The authentic aesthetic quality of this industrial product was in particular demand from the avant-garde and those architects, who wished to gain new architectonic expression from the industrialisation of building. From the time of its invention, the material has successfully performed the balancing act between imitation and originality. Today we can see the material has two faces: it is both a replacement product with a slate-like texture or wood grain, and it is an unmistakable industrial product with its own identity. Berlin architect Konrad Wohlhage describes these properties of fibre cement as "an optically perceived force that comes from within: through the depth of its surface, through its texture, through its velvetiness, which can change quickly with the light and weather." [2] So today fibre cement not only has a technological history, it also has its own place in the history of archi-

Deutsche Oper opera house Fly Tower, Berlin, 1961, Werner Düttmann

tecture and design. Werner Oechslin, Chairman of the Institute of History and Theory of Architecture at the Swiss Federal Institute of Technology (ETH) in Zurich, confirms, "undoubtedly there is a culture associated with the one hundred year history of Eternit." [3]

Fibre cement soon became the epitome of modernity and it is hard to imagine architecture today without it. In the 1920s, people experimented a great deal with the material but it saw its greatest boom in the 1950s and 60s. Renowned architects of all generations have used fibre cement products and have contributed to product development.
Le Corbusier paid a great deal of attention to the material. As early as 1912 he chose fibre cement sheets for cladding the roof on his parents' villa in La Chaux-de-Fonds, which gained the name "Maison Blanche" because of its white rendered facade. He also used a large format fibre cement panel for the front door. Shortly after this he discovered its suitability for furniture and designed cupboards out of fibre cement. For a while Le Corbusier was also engaged in product development at Eternit AG, Switzerland, and designed complete kits for prefabricated houses from fibre cement. The most comprehensive use of fibre cement products by Le Corbusier took place 45 years after his first encounter with Eternit: in his "Unité d'habitation", built as part of the 1957 International Building Exhibition in Berlin, he used fibre cement for facades and balconies, sun blinds, stair balustrades, floor coverings, ceiling linings, heater cladding, bath panels, window sills, meter panels, refuse ducts and installation pipe work. The characteristic fibre cement spindles are still standing outside the building today as ashtrays.
Walter Gropius, the Director of the Bauhaus in Dessau, also worked with fibre cement: in Stuttgart he built two similar houses using different types of construction as part of an exhibition estate, the Weißenhofsiedlung. One house was built traditionally of masonry and render. The other was built completely in dry construction – with walls of fibre cement sheets. Gropius recognised considerable advantages in this latter form of construction: no trapped moisture, a higher degree of prefabrication and a shorter construction time.
Fibre cement, which was generally referred to by the generic name "Eternit", found some prominent proponents. Adolf Behne had this to say in 1923 in his influential publication on functional buildings, "Der moderne Zweckbau": "We must turn to new construction materials: steel, reinforced concrete and in particular, Duralumin, glass and Eternit." [4]
In 1926, Hannes Mayer, who later became the Bauhaus Director, recommended "We should use synthetic rubber, wood-metal composites, reconstituted wood, plywood, rubber, glass concrete and Eternit in our buildings." [5] And 50 years later, Swiss writer Max Frisch enthused about his earlier career as an architect in his 1975 autobiographical novel "Montauk": "Cement, Sika, clinker, glass wool, Eternit. They are the vocabulary of my calligraphy." [6]

Unité d'Habitation, Berlin, 1957, Le Corbusier

Single-family house and studio of architect Egon Eiermann, 1962, Baden-Baden

Admittedly the new industrial construction material was not always welcome everywhere. Heritage protection professionals had reservations about the grey cement material. The Association of Swiss Architects produced a report in 1916 that detailed what it considered the positive and negative areas of application of fibre cement. It supported the material's use for industrial and commercial buildings but rejected the recladding of old tile, thatch or slate roofs with large, rectangular fibre cement sheets in villages and farmyards. However, over the years, fibre cement products have managed to establish themselves even in refurbishment projects and have been used in the renovation of prominent listed buildings such as the Postsparkasse in Vienna, where Otto Wagner specified it for the roof parapet.

Many buildings with fibre cement roofs or facades are now in their turn listed, and not just in Europe. An outstanding example in Japan is the 1957 home of the Japanese architect Kenzo Tange in Tokyo. He combines traditional Japanese motifs and construction with modern cement-based materials. He designed panels of fibre cement in the facade of the wooden house to slide to the side and negotiate directly between internal and external space, a function performed earlier by thin paper walls.

This Japanese lightness was an early influence on German architect Egon Eiermann. He saw fibre cement as a suitable material for contemporary interpretation. Like no other, Eiermann has shaped modern post-war architecture in Germany. And second to none he has made full use of fibre cement in his work: a material that has lent its unmistakable expression to a new modesty in German architecture. Functionality, modernity and appropriateness are the terms that Eiermann associates with Eternit. The architect, born in 1904, successfully distanced himself from the megalomania of national-socialist power architecture and designed buildings with modern appeal and expression in the 1930s and 40s. In the Berlin residential suburb of Grunewald he built a house between 1938-42 that echoes Japanese architecture. The single-storey building is divided into several wings. Exposed yellow brickwork and Solnhof limestone floor tiles stand in exciting contrast to the slightly inclined roof of corrugated fibre cement sheets. Particularly in this choice of materials, the attraction of the new can be felt. In the post-war period, Egon Eiermann used corrugated fibre cement sheets on a large scale for a handkerchief weaving mill completed in 1949/50 in Blumberg. The 112-metre long and 50-metre wide production building had to meet some very particular requirements: the loom rooms required a building with a constantly high room temperature and a relative air humidity of at least 70%. Corrugated fibre cement sheets, with their resistance to moisture and vapour permeability, offered the optimum physical properties. Eiermann used them for the roof and the facade to create a building skin that looks as if it were made in a single mould, and it became an icon of industrial architecture. The corrugated profile of the walls and gable – and also the suspended, hovering canopy of corrugated sheets – makes reference to the textile content of the production facility. Architecture critic Wolfgang Pehnt calls the handkerchief weaving mill "a revelation of modern possibilities, designed with a light hand and in no way the product of a formula." [7]

Eiermann remained true to the material and, with his own house in Baden-Baden which was completed between 1959-62, he took the use of fibre cement to a new height. A slightly inclined, projecting roof of corrugated fibre cement sheets without gutters gives the house the appearance of a completely glazed building. The roof edges are elegantly stepped with a shorter corrugated sheet.

However, Eiermann did not see fibre cement as a material only for houses or commercial architecture. Eiermann also chose the material for the young federal republic's most prestigious building: the German Parliament building in Bonn, built between 1965-69. Where others would possibly have used natural stone, Eiermann quite naturally used flat facade sheets of fibre cement. As part of this facade system he developed laminated elements for the balustrades. Through reduction and articulation, design and materials Eiermann avoided any expression of might and this was his contribution to democratic building. The house was listed in 1997.

Prototype Expo-pavilion, Montreal, 1967, Frei Otto

Eternit factory, Heidelberg, 1954, Ernst Neufert

Other prominent architects who used fibre cement in Germany in their designs to great effect after the second world war include Günter Behnisch, most notably in the construction of schools, and Otto Steidle, most notably in residential buildings. Behnisch also chose fibre cement panels for the coloured facade bays in the glass facade of the Landesgirokasse bank building in Stuttgart in 1997. A reserved and discreet material, fibre cement became increasingly popular for cultural buildings. The German engineer and architect Frei Otto clad the expressive suspended roof of the prototype of the German Pavilion for the World Exposition in Montreal in 1967 with fibre cement slates. The Austrian architect Gustav Peichl also used roof sheets on the Austrian Pavilion at the World Exposition in New York in 1964.

Four residential buildings incorporating fibre cement have become milestones of modern and post-modern architecture: the 1935 single family house by Le Corbusier in Les Matthes, southern France, the 1949 Case Study House No. 08 by Ray and Charles Eames in California, the 1960s Haus Lieb by Robert Venturi and the 1980s house by Frank O. Gehry in Santa Monica. The material experienced a high point in 1987 when it was taken up by the new architectural avant-garde and used for the warehouse of the Swiss herb candy manufacturer, Ricola, in Laufen by architects Herzog & de Meuron. The slanting fibre cement boarding articulates and provides rhythm to the facade of a simple functional building with which the architects celebrate the simplicity and potential for expression in design offered by the material.

Fibre cement did not remain solely a material for "builders of boxes", but found its place in expressive high-rises such as the student hall of residence built by Coop Himmelb(l)au at the historic Gasometer site in Vienna in 2001, and in sculptural office buildings, such as the 2005 Caltrans Headquarters built by Pritzker Prize winner Thom Mayne from the USA. Young architects are increasingly rediscovering the material, and valuing in particular its universality for use in roofs, facades and interiors. In 2001 the Dutch architects MVRDV used black corrugated fibre cement

Eternit Guesthouse Berlin-Grunewald, 1955, Paul Baumgarten

sheets to completely clad the roof and facade of a single-family house in the Hageneiland residential development in Ypenburg, the Netherlands: an approach that is being increasingly adopted in other countries too. Willem Jan Neutelings used large format fibre cement sheets and small format shingles in elegant combinations with wood and masonry in dwellings in the Netherlands. In Asia as well, the weather resistance of fibre cement in regions with high humidity means it can play an important role, especially in housing construction. In 2004, the new airport in Algiers was constructed with a fibre cement facade.

Long before corporate design and corporate architecture were introduced as an important element in modern marketing, manufacturing companies recognised that an important element of their company culture was to be found in the building culture.
In Austria Fritz Hatschek, the nephew of the inventor, reflects that the material "has had a good friend in the architect Clemens Holzmeister." [8] Even for a simple functional building like the water tower in the Eternit factory compound in Vöcklabruck, a well-known architect at the time, Mauriz Balzarek, a pupil of Otto Wagner, was engaged. Architects also designed exhibition and trade-fair buildings using fibre cement. Particularly striking were the "Wohnröhre" (Home Pipes) designed by architect Hans Hollein for the 1969 International Water Supply Congress in Vienna. And from an early date, the company involved itself with cultural and social projects in which great store was set by the architectural quality of the material. Architects and former Wagner pupils Heinrich Schmid and Hermann Aichinger, who had become known in particular for their residential and community buildings, were commissioned in 1927 with the construction of the Hatschek Hospital in Vöcklabruck.
The architectural tradition of Eternit Switzerland is even more characteristically evident. The sales office at Eternit AG in Zürich, which was built in 1956, was the work of architect Otto Glaus, the factory in Payerne was designed in the same year by Paul Waltenspühl, the research laboratory in Niederurnen was completed in 1959 by Thomas Schmid, and the administration building by Haefeli, Moser, Steiger in 1954. This lively architectural tradition continued with the painstaking modernisation of historic structures. The Zurich architect Stefan Cadosch is the current designer of room and trade-fair architecture for Eternit in Switzerland.
The record shows a similar continuity in Germany. In Berlin Paul Baumgarten was the company's customary architect for many years. He was responsible for the design of the sales office in Berlin's Tiergarten district in 1957, which today is a listed building and is known as the "Eternit-Haus". His other works included the cafeteria on the former works site in Berlin-Rudow and numerous factory sheds. The company's cultural identity includes its own hotel in Berlin-Grunewald, also designed by Paul Baumgarten in 1954. Architect

Gasometer, Vienna, 2001, Coop Himmelb(l)au

Ernst Neufert led construction projects at the southern German branch of the company in Heidelberg between 1954 and 1964. Neufert, who gained international fame as the author of the "Architects' Data" manual, also wrote two important construction manuals for Eternit, which have been revised and reprinted eight times. Neufert was the architect responsible for all the Eternit factory sheds with their striking facades and shed roofs made of corrugated sheets, and the relatively modest administration building: an uncluttered, linear 1960s box. This building reflects the architectonic aspirations and understatement of Modernism. This architectural tradition is also continuing. The design for the modernisation of the facade and internal rooms of the administration building was given to the young Berlin architect Astrid Bornheim.

Ricola Storage Building, Laufen, 1987, Jacques Herzog & Pierre de Meuron

A rather special feature is the international architecture magazine about fibre cement published on behalf of Eternit AG. The publication appeared from 1956 to 1984 as "ac revue" and since 1993 has continued under the title "A+D Architecture and Detail – Building with Fibre Cement". In 1957 Sigfried Giedion, the first publisher of the "ac revue", wished to have a house in the mountains, and because he thought it was prudent not to confuse his career as an architectural historian with that of an architect he sought eminent advice from Konrad Wachsmann and Le Corbusier. Both designs were unacceptable to the natural heritage conservation authorities. In the end he chose Eduard Neuenschwander as his architect and opted for an external wall cladding of fibre cement panels "in a brownish red colour, as used in paintings by Picasso, Braque or Corbusier." The panels were designed as a 33cm wide module "... carefully fastened in place, the sheets and their colour have proved a thorough success in the landscape. For it is by no means the case that industrial prefabricated building materials cannot be made to fit in with a new regionalism." [9]
Fibre cement is a material that does not allow itself to be co-opted in the interests of any particular cause – either for regional architecture trends or for particular types of building. The material is visibly just as resistant to rainstorms, frost and fire as it is to fashions, style and contemporary taste.
Independence, neutrality and universality ensure that the material is still interesting to architecture and design, even after 100 years or more since its invention. The fibre cement beach chair "Loop" by the Swiss designer Willy Guhl in 1954 has since become a design classic. Architects and designers are taking renewed interest in unusual applications of the material – whether that is as the washstand by Astrid Bornheim or as the office luminaire by Rupert Kopp of Greige Design. Product development follows on from successful cooperation with architects. Consequently over

the years Arno Lederer in Stuttgart, Rainer Hascher, Armand Grüntuch and Konrad Wohlhage in Berlin, David Trottin of Peripheriques Architectes in Paris and Ash Sakula Architects in London have all been involved.

In art as well, there is a long line of tradition: from Pablo Picasso's paintings on fibre cement through the fibre cement sculptures by Bernhard Heiliger to the photographic series of materials, products and production processes by Karin Geiger or Folke Hanfeld. The versatility and formability of the material offers still more scope for development, invention and experiment, enabling the claim contained in the name to be taken up again in new ways: in this respect Eternit should be associated more with "timelessness" than with "eternity".

Experimental design in fibre cement, Hautlabor, 2004, Astrid Bornheim

Technology

The manufacturing process for fibre cement has remained the same for more than 100 years. Only the composition of the material has changed: natural asbestos reinforcing fibres have been replaced by a mixture of environmentally and health-compatible synthetic and cellulose fibres. These high performance fibres are responsible for the high strength in tension and bending of the thin cement sheets. This allows fibre cement sheets just 8mm thick in formats up to 3100 x 1500mm to be used for ventilated cladding facades on all types and heights of buildings. 4mm thick slates with simple fastenings have a proven record of success in pitched roofs. Traditional formats such as rhomboid or square diagonal lap stand alongside modern cladding patterns like rectangular common lap. The low mass per unit area of fibre cement slates also makes them suitable for the refurbishment of the roofs of historic buildings. The special shape of corrugated sheets means that their supporting members can be placed at greater distances apart and makes them particularly suitable for large shed roofs. A huge range of preformed pieces and make-up sheets ensures precision in detail for design and construction. New through-coloured sheets and slates, new coating technologies fine-tuned for the material and optimised fastening systems are currently the most significant product developments. In addition to their range of high-tech mass produced standard products, manufacturers can also supply preformed pieces, furniture and plant containers, as well as hand-crafted items with free-form shapes and complex geometries.

Fibre cement is a material made from cement and synthetic fibres. Fibre cement panels are used as small format roof and facade slates, corrugated sheets and large format sheets for facades and internal finishings. They can also be handmade to any shape for complex corrugated sheet geometries, plant containers or design objects. Experience with this technology has been amassed over more than 25 years: development, observation and experience in laboratory and accelerated tests, as well as many years of exposure under real conditions in built projects. Fibre cement is non-combustible, dimensionally stable and resistant to weather. Its chemical resistance and ageing behaviour are comparable with those of concrete. The main raw ingredient is its binder, Portland cement, which is produced by burning limestone and clay marl. Additions such as limestone dust are generally added to improve product properties. Synthetic organic fibres of polyvinyl alcohol are used as reinforcement. These fibres are used in a similar form in textiles, protective fabrics, non-woven fabrics and surgical stitches. During the manufacture of fibre cement, process fibres act as filter fibres. They are mainly cellulose fibres, similar to those used in the papermaking industry. Fibre cement also contains air in the form of microscopic pores. A system of micropores creates a frost-resistant, moisture-regulating, breathable yet waterproof construction material. Products made from fibre cement are neutral with respect to electromagnetic radiation, which means that signals from pagers, radio, infrared and radar equipment pass through unhindered.

Manufacturing process

Fibre cement products can be manufactured as normally cured, air-dried panels, or as autoclaved, steam-cured sheets. In the Hatschek machine, which was named after its inventor, Ludwig Hatschek, a water-based slurry of fibres with adhering fine cement particles is formed into a thin layer on the surface of a sieve cylinder. Three such sieve cylinders rotate with up to two-thirds of their diameter immersed in the slurry. The parts of the sieve cylinders projecting out of the slurry come into contact with an endless, continuously moving felt belt. This felt takes up the fibres and the adhering cement particles to form a thin layer of randomly oriented fibres. The still very moist fibre layer is transported on the rotating transport felt, has its water reduced by a suction system and is then transferred to a forming roller. This process continues until the required thickness is achieved. The fibre cement jacket is cut to length and the uncured sheet is transferred to a cutting and punching press. The large format sheets are stacked between metal plates and the stack is placed in a press to further compress the material.

After the corrugated sheets, which are still moist at this stage, have been cut to the specified size in the press, they are formed into the final corrugated shape using a suitable corrugator in the suction plate system and placed between corrugated moulds. The sheets remain in temporary storage to allow the cement to cure. As soon as it has reached an adequate green strength for the corrugated sheet to be removed from the stack, it is separated from the mould. After three weeks of temporary storage, a high quality, durable and weather-resistant coating is applied to the corrugated sheets. Roof slates and facade sheets receive similar surface treatment and a coloured or transparent coating is applied.

In addition to the mass production of fibre cement products on virtually fully automated production lines, standard preformed pieces and customised products can be hand-moulded in traditional ways. The moist fibre cement layer is placed over moulds and given its final shape by specialist craftsmen. Some preformed pieces from the corrugated sheet production line are cut after curing and stuck together with special fibre cement adhesive. It is only cost-effective to spray mould fibre cement if large batches are to be produced. This spray-moulding process is usually only used for mass-produced products such as the two-piece ridges with corrugated wings.

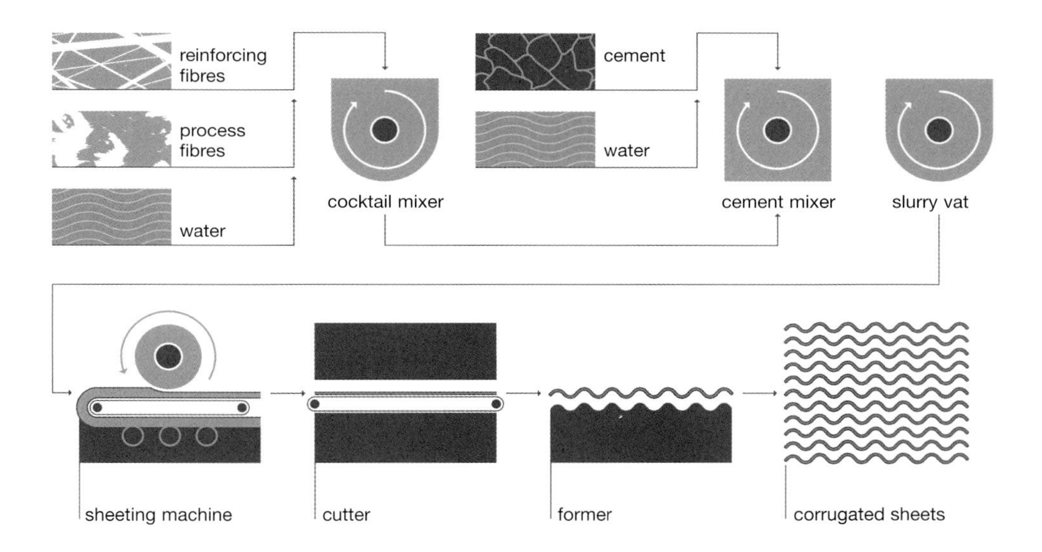

Manufacturing process fibre cement corrugated sheets

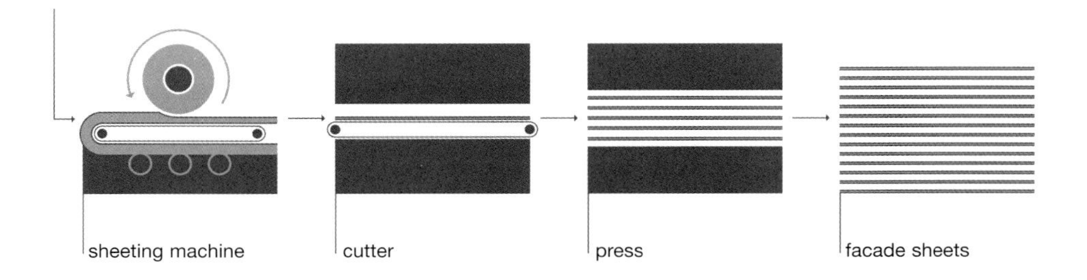

Manufacturing process fibre cement sheets

In this chapter, drawings are not according to scale

LARGE FORMAT FACADE SHEETS

Large format fibre cement sheets for facades and interiors can have a maximum effective size of 3100 x 1500mm. The maximum size depends on the width and diameter of the forming roller. Various lengths and widths are available, depending on the product and the manufacturer. The sheets can normally be produced in thicknesses between 6 and 20mm. The usual thicknesses for facade sheets are 8, 12 and 15mm, depending on the fastening system and the applicable national construction regulations. Typical weights of fibre cement products are 15.4kg/m² for 8mm thick sheets, 22.8kg/m² for 12mm thick sheets and 28.5kg/m² for 15mm thick sheets.

Areas of application

Large format fibre cement sheets are primarily used for:

- ventilated external wall cladding
- cladding for post-and-rail construction
- weatherboarding
- external cladding of prefabricated composite laminated elements
- balcony slabs
- window reveals
- window and door lintels
- internal walls

Construction principle of ventilated cladding facades

The most commonly used construction of a fibre cement facade follows the principle of the ventilated cladding facade. The whole construction is open to vapour diffusion. The air gap prevents moisture damage and heat from building up. Any thickness of insulation can be incorporated. Installation is not affected by the weather. The ability to accommodate construction tolerances is a big advantage especially in refurbishment projects. All components of the aluminium or wooden substructures, insulation and cladding elements can be separated and recycled.

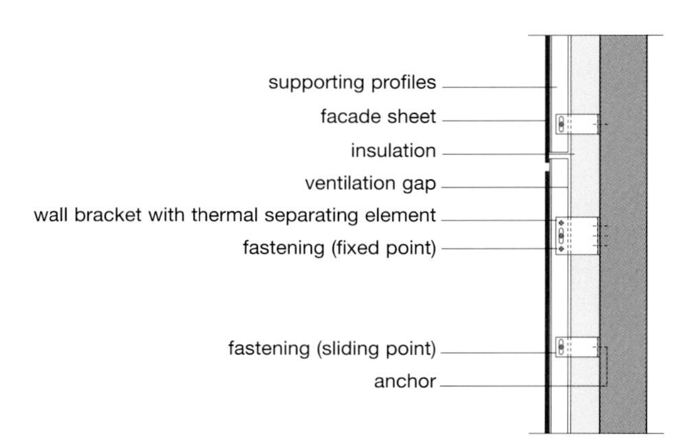

Principle of construction of ventilated cladding facade

Fastening systems

The choice of fastenings is an important decision in the design of a fibre cement ventilated cladding facade. It is important both in terms of appearance and economy. In addition to questions of construction-authority approval and cost, there are also technical requirements, such as fire protection, to be investigated and complied with. Across the whole range of available materials and systems, fastening techniques can be grouped into two basic types: visible and concealed fastenings. Wood and aluminium substructures have proved their worth in use. Wooden substructures are generally suitable for anything smaller than a high-rise building. Aluminium substructures can be used for any type or height of building.

Surfaces and colours

The natural, untreated surface of fibre cement is velvety and open-pored. If water penetrates fibre cement it can lead to efflorescence. Dust and dirt marks may become established. Therefore, the use of coated fibre cement panels is recommended for areas where the facades and internal walls are left open to view. It is important that the rear side is sealed in the same way with a physically equivalent surface treatment to the front in order to equalise internal stresses in the panel. Coatings applied in the factory are generally pure acrylate. The coloured pigments used must be UV- and alkali-resistant to ensure long-term colour stability.
Fibre cement sheets can be through-coloured by incorporating pigments in the slurry. Matt, transparent or coloured varnishes further enhance the material's aesthetic quality whilst allowing the character of the uncoated sheets to express itself. Slight irregularities, differences in colour and marks from the manufacturing process are typical. A coloured covering coat with a strong colour accentuation is another option. Impregnation of the edges with a temporary transparent medium is required on varnished and through-coloured sheets to prevent water penetration at the sheet edges. Otherwise, in wet weather the take-up of moisture at the edges of the sheets can make them appear a darker colour. In dry periods of weather these dark areas may disappear, making them a cyclic effect dependent on the time of year. Impregnation considerably reduces this water take-up. The impregnation must remain effective until the carbonation of the sheet is complete and the sheet has fully cured. It is not necessary to impregnate the edges of the drilled holes.

Cutting fibre cement sheets

Large format fibre cement sheets are normally cut to the correct length and preconfigured for the fastenings at the factory so that any necessary cutting on site is limited to a few fitting cuts. Diamond or hardened metal-tipped saw blades are the most suitable for cutting fibre cement sheets on factory cutting machines. Optimum cutting speeds are 60m/s with diamond-tipped cutting machines, 2-2.5m/s with hardened metal-tipped saws. The ideal material feed speed on diamond-tipped cutting machines is 20m/min, on hardened metal-tipped saws 3.0-3.5m/min. The fibre cement panels should be sawn with the colour coating facing upwards to produce the best quality of cut. After cutting, the edges are lightly abraded. Silicon carbide grinding discs and diamond cutting discs are not to be used for cutting fibre cement products. This advice applies to dry and wet cutting. The reason is that both these types of disc cutters operate at high rotational speeds and the associated high cutting pressures can lead to excessive loads on the material at the cut edges. Another method of cutting fibre cement sheets is the water jet process. This is of particular interest if non-rectangular or irregular shapes are required. Make-up pieces can be cut on site

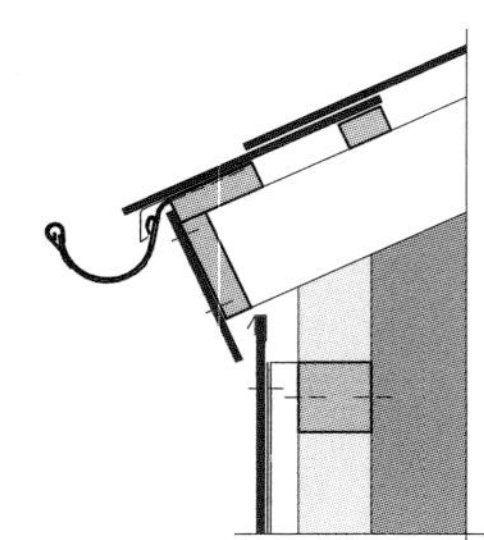

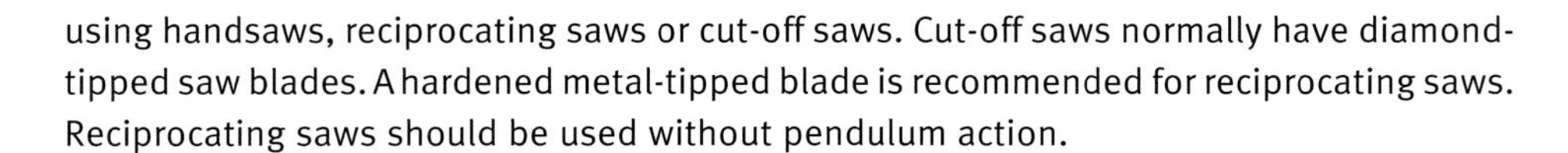

using handsaws, reciprocating saws or cut-off saws. Cut-off saws normally have diamond-tipped saw blades. A hardened metal-tipped blade is recommended for reciprocating saws. Reciprocating saws should be used without pendulum action.

Transport and storage

Fibre cement facade sheets should be stored and transported on a level, dry supporting surface and be in full contact with that surface. They should be stacked alternately visible face to visible face and back to back. The visible faces should be separated by release paper to prevent scratches. The sheets should be protected against moisture and dirt by covering them with construction foil until installation. They should not be pulled off the stack but rather lifted and always carried vertically on one end.

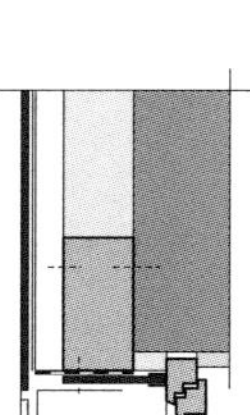

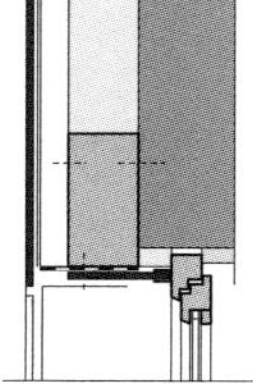

Fastening to wooden substructures

The fastening system must not constrain movement of the sheets. Stresses arising from constrained changes of shape must not be allowed to cause damage to the cladding or the substructure at the fastening points. Drilling all the fastening holes in the sheets 2mm greater than the shaft diameter of the facade screw will ensure that the sheets are not constrained when mounted on a wooden substructure. The minimum edge distances of 80mm in the direction of the supporting wooden battens and 20mm transverse to them must always be observed. The maximum edge distance of 160mm should not be exceeded. Edge distances up to 200mm are permissible in special cases, e.g. over window blind boxes. Edge distances greater than 160mm may lead to slight differences between the planes of the surfaces of adjacent sheets. This has no detrimental effect on the structural stability of the facade. There should be a joint sealing strip of adequate width between the sheets and supporting battens to protect the wooden substructure from moisture damage. These construction measures will prevent long-term moisture saturation of the battens. The EPDM or bitumen-coated aluminium must extend at least 5mm beyond the edges of the battens. The width of the joints between the large format fibre cement facade sheets is normally 10mm. Joints less than 8mm wide must be avoided. Open joints wider than 12mm should be avoided. If the horizontal joints are left open this will considerably reduce the amount of dirt accumulating on the facade. The additional cross-sectional area in the ventilation openings, created in this way contributes greatly to the efficient functioning of the ventilated cladding facade. A facade with open joints (8-10mm) still provides full rain protection. If joint profiles are used then they should not be more than 0.8mm thick. They can be kept from migrating by simply fixing them in place. Horizontal joint profiles will increase the accumulation of dirt on the facade. Overlapping joint profiles are to be avoided. Corner profiles should not constrain the movement of the facade sheets. Free movement joints must be provided.

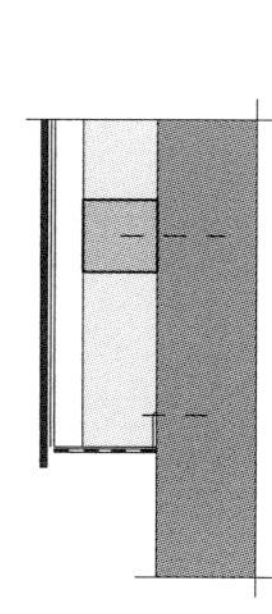

Fibre cement facade on a wooden substructure – standard details: eaves, lintel, balustrade, plinth

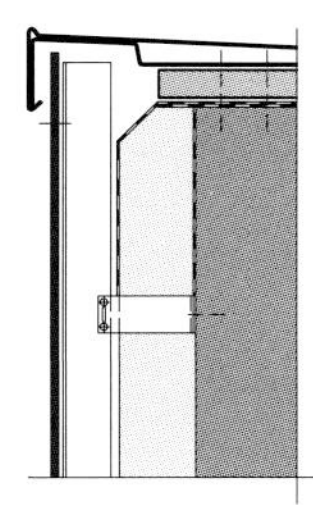

Fastening to metal substructures

The cladding in the area of a movement joint must be able to accommodate the same movements as the joint. So that no movement constraint can occur as a result of joining individual sheets by aluminium vertical supporting profiles, these profiles should not be butt-jointed between the fastening points of a sheet. Otherwise this joining of individual sheets by aluminium supporting profiles leads to constraints or excessive movements that in turn result in damage. The supporting profiles must be positioned and aligned in the substructure in such a way that the facade sheets all lie on the same plane and can be fastened in place without constraint. The fixed points of the supporting profiles to which a sheet is fastened must be at the same height. It follows, for example, that at window balustrades the profiles must be made discontinuous to prevent profiles from butting up behind the sheets.

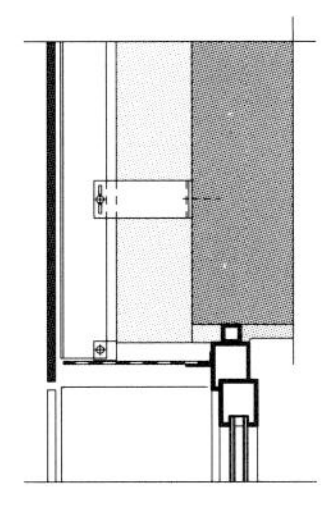

Riveted fastenings

A clean pattern of fastenings is guaranteed by predrilling holes in the sheets to millimetre accuracy. For riveted fastenings with 4mm rivets all holes must be predrilled with 9.5mm diameter holes. Drilled facade sheets are fastened to a metal substructure at fixed and sliding fastening points. There are two fixed points per facade sheet, which are formed with fixed point sleeves. They guarantee the accurate and stress-free support of a sheet on the substructure. The minimum edge distances of 80mm in the direction of the supporting profiles and 30mm transverse to them must always be observed. The maximum edge distance of 160mm should not be exceeded. Edge distances of up to 200mm are permissible in special cases, e.g. over window blind boxes. Edge distances greater than 160mm may lead to slight differences between the planes of the surfaces of adjacent sheets. This has no detrimental effect on the structural stability of the facade. The use of bitumen coated aluminium supporting profiles prevents undesirable reflections in the joints. There must never be two fixed points on the same substructure profile. This means that on each sheet there is always one fixed point support at right angles to the direction of span of the supporting profiles. The two fixed points must be placed as centrally as possible in the sheet. If possible each fixed point is to be installed on the second profile in from the right and left edges of the sheet. If the joints are to be highlighted, joint cover plates made from coated aluminium can be used. They are to be no thicker than 0.8mm. The joint profiles are not to be overlapped at the intersection points. The joint profiles can be black or matched in colour with the facade sheets. If horizontal joint profiles are used then an increased and uneven accumulation of dirt on the external facade surface can be expected. Exposed aluminium parts for use in facades must be coated. Uncoated aluminium may discolour unevenly and this can lead to distracting dirty marks on the cladding.

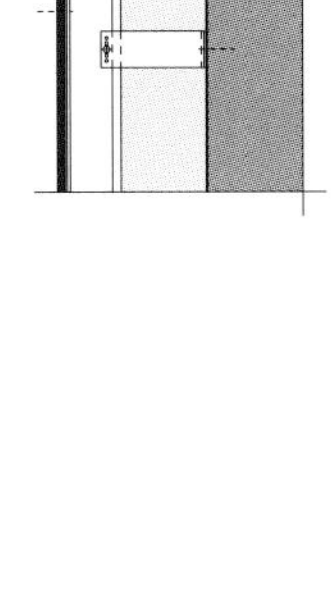

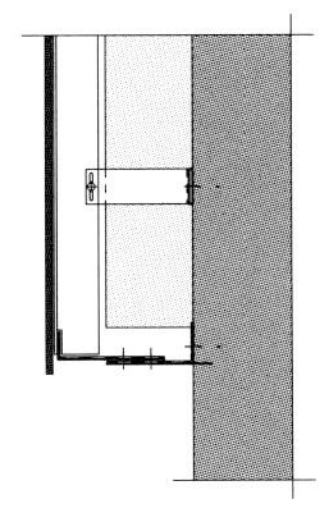

Fibre cement facade on an aluminium substructure – standard details: eaves, lintel, balustrade, plinth

Standard details

The examples of standard details shown here are frequently used in practice. In the sketches of aluminium substructures the support profiles have been simplified as angle or T-profiles. The connection to the wall depends on the type of substructure system. Ventilated external wall cladding must be provided with ventilation openings with a minimum open cross-sectional area of 50cm² per 1m length of wall. To prevent entry by vermin and birds the ventilation openings should be covered with perforated profiles. The ventilation opening area should not be less than 40% of the leg area. To prevent stresses arising in the cladding the ventilation profile should be fastened to the external wall. If this cannot be achieved for constructional reasons and a ventilation profile has to be fastened to the support battens or supporting profiles then its thickness should not exceed 0.8mm. If the thickness exceeds 0.8mm then the ventilation sheet must be mounted behind the supporting profile.

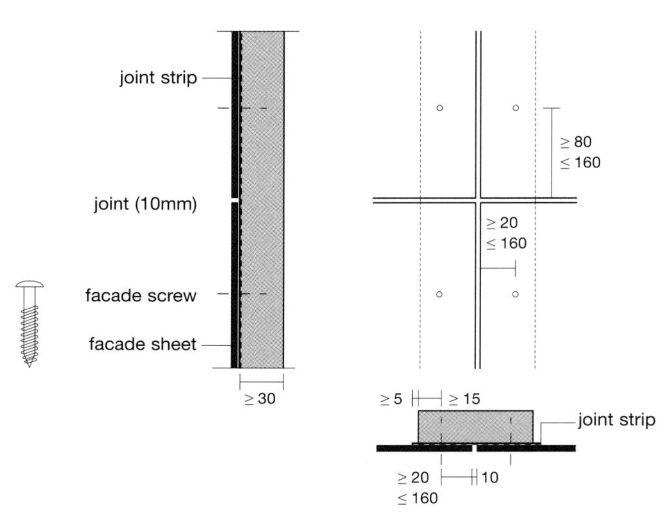

Fibre cement facade on wooden substructure

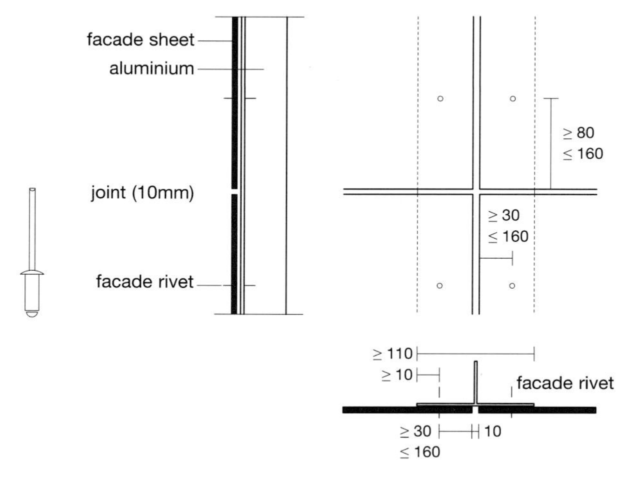

Fibre cement facade on aluminium substructure

Rear fastenings

Facade sheets can also be fastened on to aluminium substructures from the rear. Installation of fibre cement sheets of 12mm thickness and greater is done by undercut anchors and agraffes or sheet supporting profiles. After inserting the anchor into the undercut hole, the leg of the anchor is brought into the desired position by tightening the screw to create a positive connection with the facade sheet. For a more secure connection with the substructure the undercut anchors are given square collars. These collars allow constraint-free connection to the components of the substructure. A punched hole can be provided here to accept the anchor collar, depending on the type of connection required: a square hole for a fixed point and a rectangular hole for a sliding point. Each facade sheet must be attached with at least four anchors in an orthogonal pattern on suitable substructures in a way that does not constrain the sheet's movement. The number of individual agraffes should not exceed nine. If more than nine fastening points are required then they must be placed on continuous sheet supporting profiles or agraffe profiles. The positioning of the holes is determined by the format of the sheets, the type of substructure, the structural engineering requirements of the facade and the edge distances of the undercut anchor holes. The edge distances must be between 50mm and 100mm horizontally and between 70mm and 100mm vertically. If edge distances of more than 100mm are used this may lead to the edges of sheets being out of plane, especially in the area of the cross joints. The appropriate agraffes for the system are attached to the rear of the sheet. The sheets prepared in the above way are then suspended from the horizontal supporting profiles, adjusted in position and made effectively and permanently secure against lateral sliding or migration with the provided clamps. The horizontal supporting profiles should have a break in them every 4m or so, in order to avoid undesirable uneven joint widths between the sheets caused by the expansion of the aluminium. The self-weight is always carried by two adjustable fastening points. The minimum construction thickness from the front edge of a 12mm thick facade sheet to the supporting surface of the wall is 100mm. Sheet supporting profiles can be used instead of agraffes. The sheet supporting profiles are attached to the rear of the sheet with undercut anchors in a way that does not constrain the movement of the sheet. After the sheets have been adjusted into their final positions the prefabricated elements are attached through the joints to the supporting profiles on the substructure. The self-weight is always carried by two fastening points.

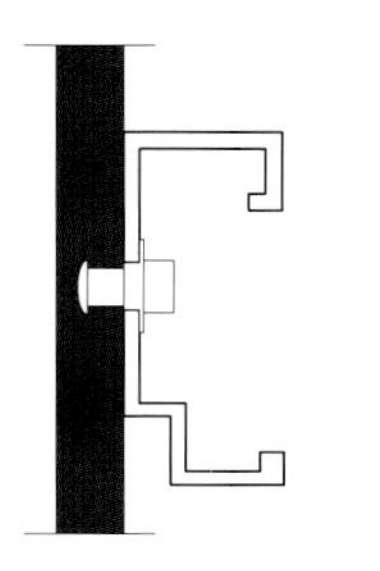

Agraffes

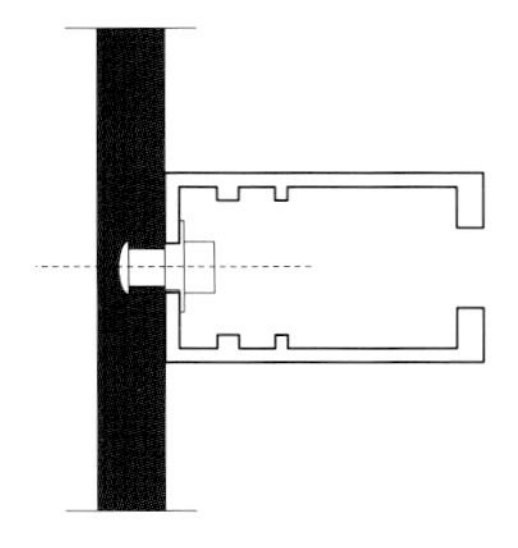

Sheet support profiles

Glued fastening systems

Facade sheets can also be attached to aluminium substructures using a glue specially developed for fibre cement. Normally a general certification for constructional use and installation by a certified specialist company are required. The glued fastening system can be used on 8mm thick sheets. The adhesive bond creates a solid connection capable of transmitting forces and therefore no other mechanical fastenings are required. Installation must be on a perpendicular aluminium substructure for ventilated cladding facades. Installation may only be undertaken under certain climatic conditions. The installation temperature must be between +5°C and +35°C. The relative air humidity must not exceed 75%. The material temperature must be › 3°C above the dew point temperature. Installation work must be protected against the weather and dust.

SMALL FORMAT FACADE SLATES

Ventilated cladding facades can also be constructed with small format facade slates. The range of sizes extends from 20 x 20cm slates for smaller vertical surfaces to 60 x 32cm slates for whole facades. Shapes, formats and placing patterns often follow regional traditions and may vary from the examples shown, depending on the region. Some of the most commonly used placing patterns are shown in the following examples, which are intended to give some insight into the basic principles and designs. The small format fibre cement slates have smooth or textured surfaces.

Areas of application
Small format fibre cement slates are primarily used for:

- gables
- dormer windows
- chimneys
- sides exposed to weather
- facades
- complete building skins

Fastenings
Facade slates are fastened with slate nails or hooks. The type and number of the fastenings depends on the slating pattern, slate size and area of application. The slates are generally placed on battens. Facade slates would only be placed on sheathing boards where small format slates are used to clad small surfaces, such as chimneys and the sides of dormer windows. Small format external wall fibre cement slating on a wooden substructure generally involves the following elements:

- facade slates/cladding
- wooden support battens or boards
- counterbattens, primary and intermediate battens or metal spacers
- connection elements
- fastening elements
- anchoring elements
- insulation, insulation fixings

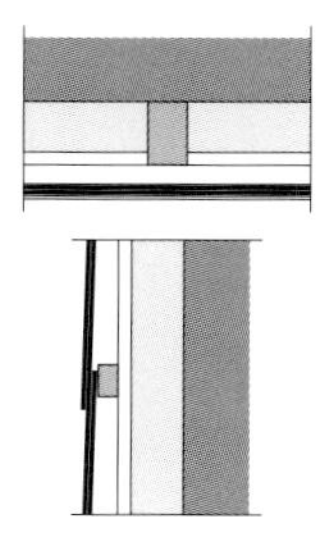

Horizontal supporting battens

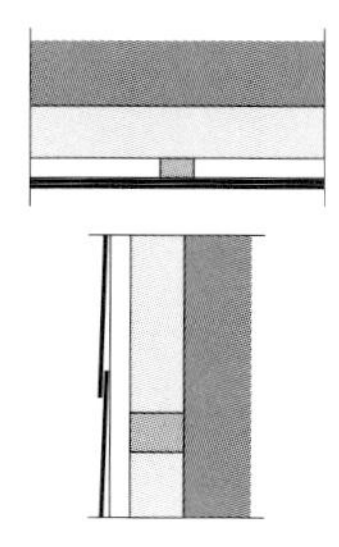

Vertical supporting battens

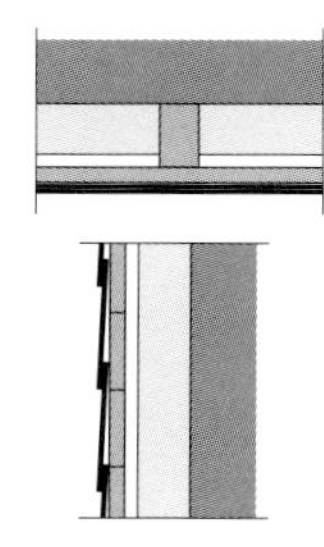

Boards

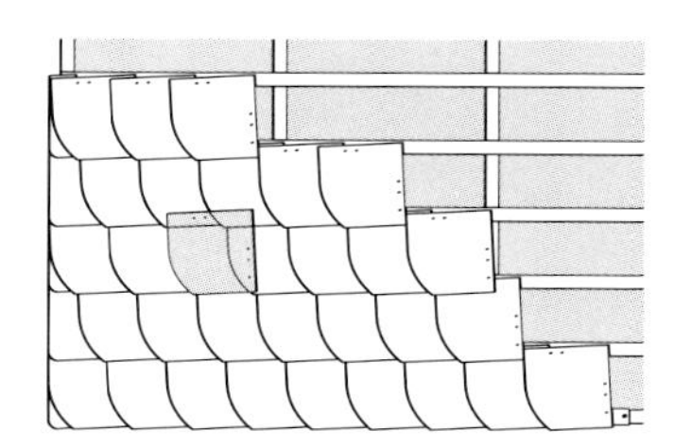

Traditional lap "Deutsche Deckung"
with curved-edge slates left,
example format 30/30cm

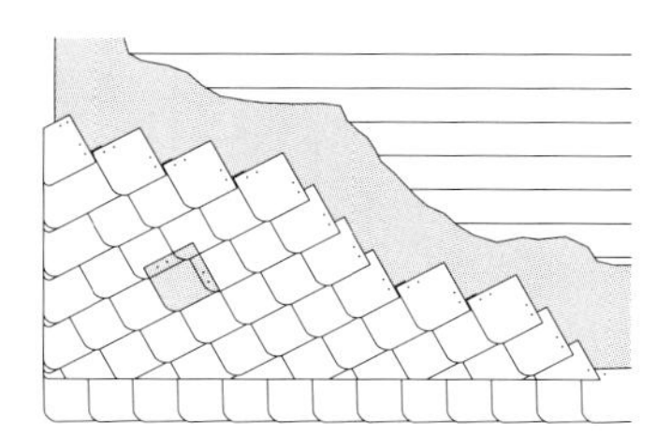

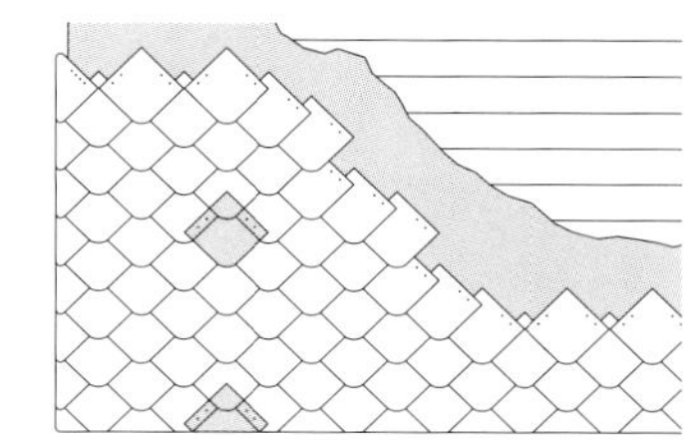

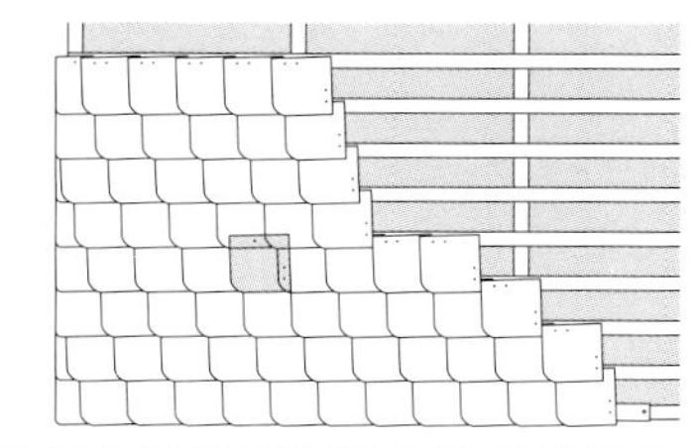

Variations of slates with rounded corner,
example format 20/20cm

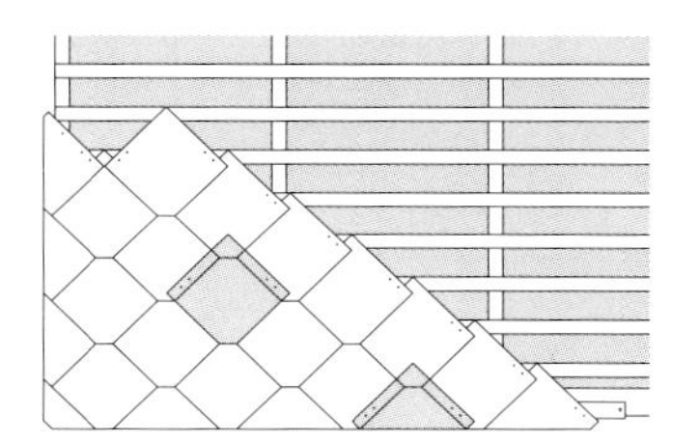

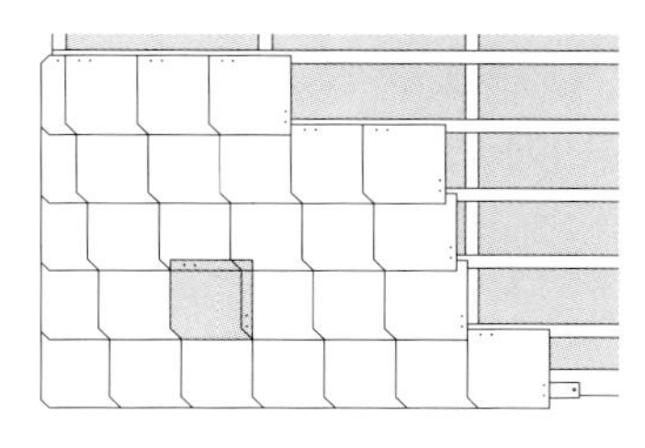

Variations of Honeycomb lap, truncated,
example format 30/30cm

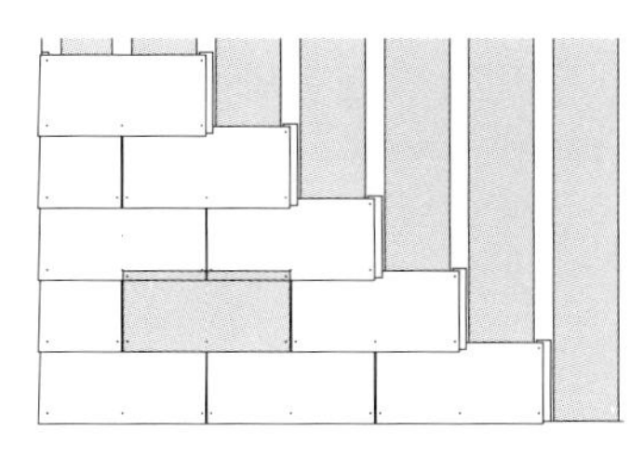

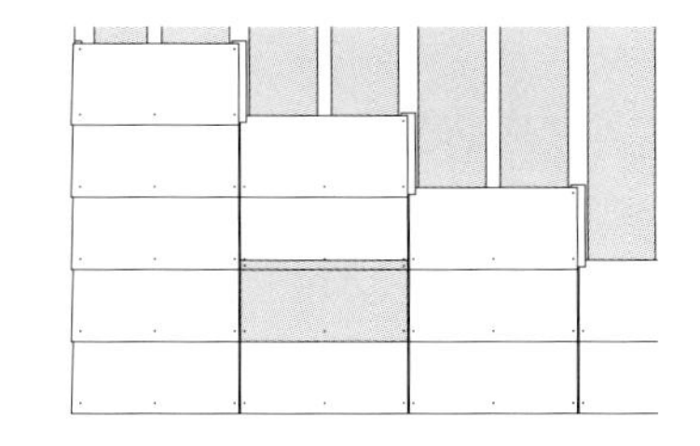

Horizontal common lap / Vertical common lap,
example format 60/30cm

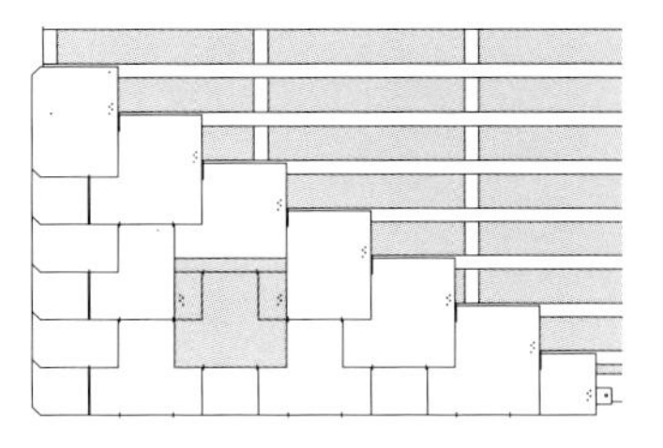

Open double lap,
example format 40/40cm

ROOF SLATES

4mm thick fibre cement roof slates can be used for linear roof shapes with flat planar surfaces. As with facade slates there are often regional variations with different shapes, formats and placing patterns. The formats and technical regulations may differ in detail from region to region and country to country. Some of the most commonly used placing patterns are shown in the following examples, which are intended to give some insight into the basic principles and designs. They can also be used for facades. Roof slates are available with smooth and textured surfaces.

Minimum roof slope

The minimum roof slope must be observed in the design and construction of sloping roofs clad with fibre cement. It is the least possible roof slope at which practice has shown that a roof cladding would still exclude rain. If a roof slopes less than the minimum roof slope additional measures are necessary to ensure it keeps rain out. If a roof is flatter than the minimum roof slope by 10° then the roof will not be effective, even with additional measures. The minimum roof slope is different for different cladding patterns:

- traditional slating "Deutsche Deckung" › 25°
- double lap › 25°
- diamond shape slating › 30°
- horizontal lap › 30°

Other cladding patterns such as open or honeycomb can only be used as external wall cladding and not for roofs.

Single and double lap

With single lap, the roof slates are lapped at the head and the side. Where the heads and sides lap, there are two roof slates covering these areas and elsewhere only one. The rainwater falling on the cladding is conducted over the edge of the overlapped slates in the direction of the eaves. Placing the cladding with the bottom of one slate resting on the head of the next lower slate means rainwater flows on to the roof slate below.
With double lap, every course of slates is overlapped, by the course above the one immediately above, by the specified head lap. Where the cladding overlaps, there are three roof cladding slates covering these areas and elsewhere two. The relatively large total head lap in comparison to the visible area of an individual slate is necessary because there is no side lap. Rainwater falling on a double lapped roof flows unobstructed down the vertical joints. Hence there must be an adequate overlap at the vertical joints.

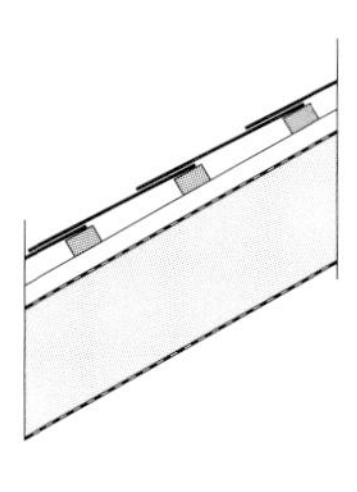

Single lap slating

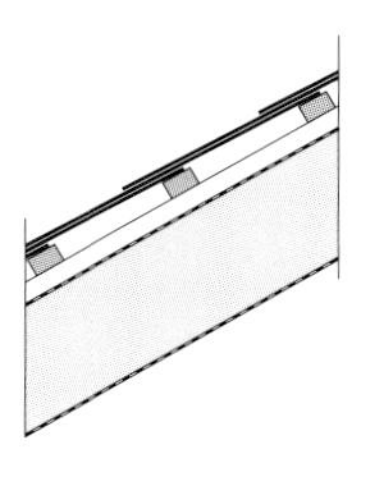

Double lap slating

Design loads

The low mass per unit area of fibre cement slates means that they are very suitable for refurbishing historic roofs without the need to renew the roof trusses. Single lapped slating including battens weighs $0.25 kN/m^2$, for double lapped slating the figure is $0.38 kN/m^2$.

Ventilation and moisture protection in roofs

Roofs with fibre cement slates can be designed and built with or without ventilation of the insulation. The height of the open ventilation cross-section in the roof must be at least 2cm. The constriction caused by fitting ventilation grills in these openings should be taken into account particularly at the eaves and ridges of mono- and double-pitch roofs. The ventilation openings may need to be enlarged to comply with the minimum opening size requirements. Roofs with thermal insulation require an effective vapour barrier placed room-side. All service feeds and penetrations must be constructed as airtight. No condensation water must be allowed to build up as a result of diffusion or convection.

Additional measures to prevent rainwater entry

Additional measures are to be taken in the design and construction where increased requirements apply to the roof cladding, such as

- the roof slope is less than the minimum
- the roof space is to be used as living space (e.g. warm roof construction)
- special climatic conditions (e.g. exposed site, frequent driving rain or blizzards)
- special construction features (e.g. large roof depth, dormer windows, roof valleys)

Fastenings

Fibre cement roof slates are fastened in place with slate nails or hooks. Larger roof slates should have at least one additional fastening. The type and number of fastenings depends on the cladding pattern, slate size and area of application. This applies to roof and wall cladding. Galvanised slate nails are used for fastening roof slates, with exception of the edge slates. On roofs with wooden components stainless steel slate nails are used. Slate hooks must be made from stainless steel or copper. Slate nails made of stainless steel or copper must have jagged shafts. The slate nails can be driven in with a light pneumatic nail gun.

Cutting and drilling

Small format roof and facade slates are normally drilled in the factory with holes for overlapping. Additional fastening holes can be punched with a special tool or drilled using a drill with standard hardened drill bits without hammer action. The holes must be between 3.5mm and 4.5mm in diameter. Cutting roof slates or trimming make-up pieces is done with special fibre cement guillotine shears, hand shears or a slater's hammer and anvil. Dust from cutting or drilling should be avoided or all traces of dust immediately removed.

Roof substructure

Roof substructures for cladding with roof slates are normally constructed in wood. In the case of roof substructures consisting of roofing membranes or underlays, and roof slates supported on battens, counterbattens are placed on the rafters to ensure rear ventilation of the roof cladding and allow any penetrating moisture to drain away.

Eaves

The roof slate cladding at the eaves must take into account the eaves construction and any required ventilation openings. Gutter supports need to be incorporated into the substructure. The roof slates at the eaves should be underlayed so that they have the same slope as the rest of the roof.

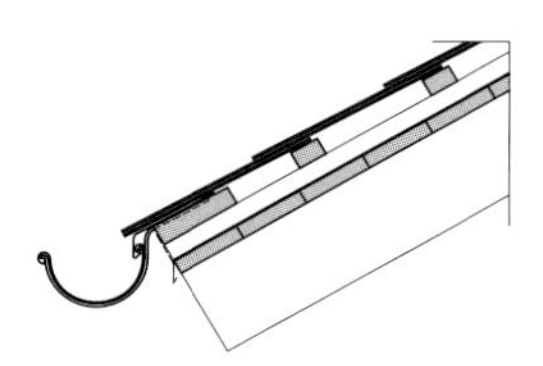

Eaves double lap on battens

Eaves single lap on sheathing boards

Double pitch roof ridge

With all types of cladding a ridge cap course is laid at the ridge. The edge slates at the gables are fastened with stainless steel or copper slate nails, which are left visible. Edge slates must be laid with a distance of at least 50cm from the edge of the verge or hip. The ridge cap course on the windward side of a double pitch roof is laid to project above the corresponding course on the leeward side. The projection must be 4-6cm. Depending on the design of the ridge it may be necessary to underlay the roof slates in the ridge cap course in order that they have the same slope as the lower slates. The ridge may be formed with a one- or two-sided metal linear ventilator. Double pitch roof ridges can also be formed with ridge capping or, in case they are side-lapped, double-lapped roof slates in the ridge cap course, with non-projecting flashings.

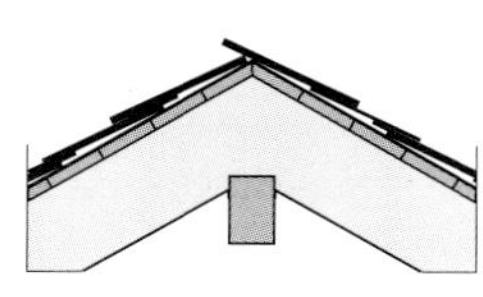

Double pitch roof ridge on sheathing boards

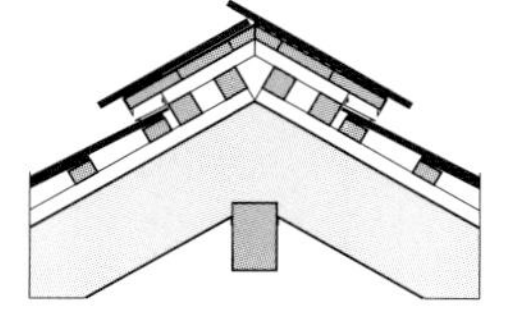

Double pitch roof ridge with vent on battens

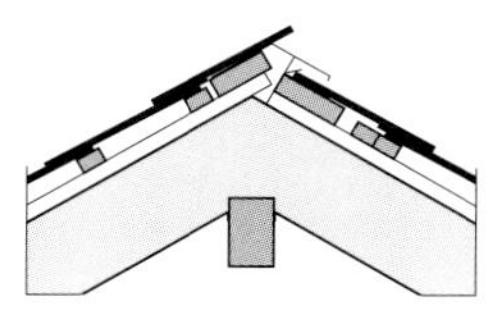

Double pitch roof ridge with one-sided linear ridge vent

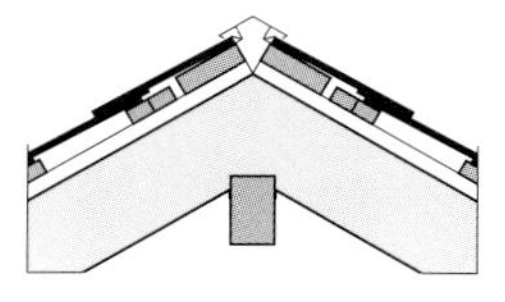

Double pitch roof ridge with two-sided linear ridge vent

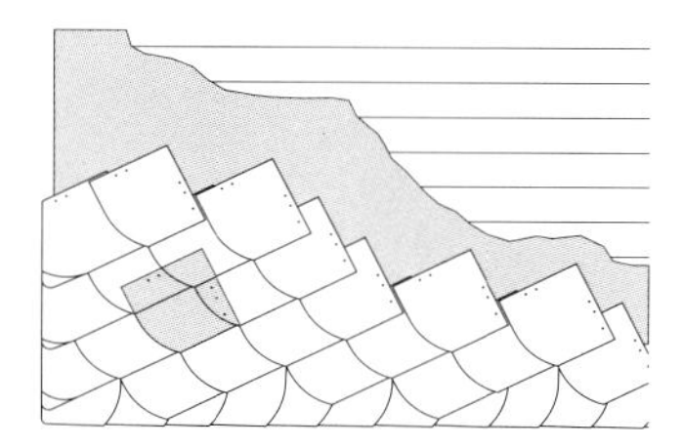

Traditional "Deutsche Deckung" with curved-edge slates and rising courses on full sheathing, Example format 30/30cm

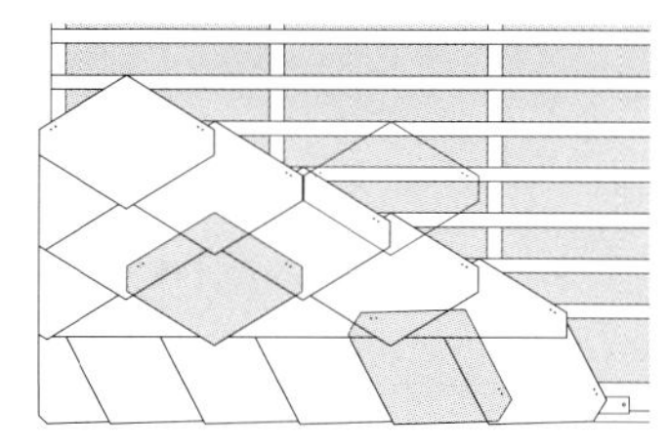

Diagonal lap (rhomboid slate), Example format 40/44cm

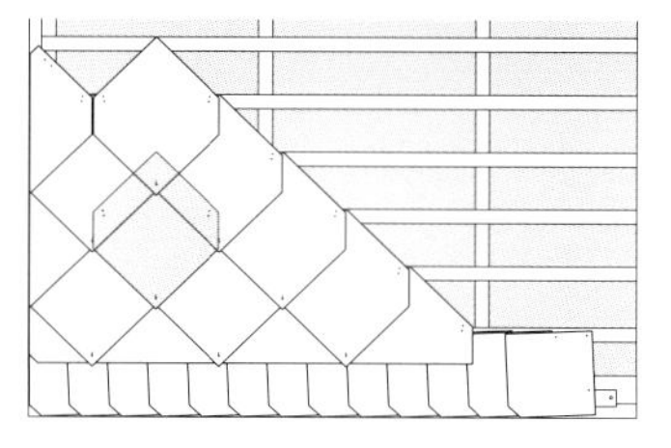

Diagonal lap (diamond shape slate), Example format 40/40cm

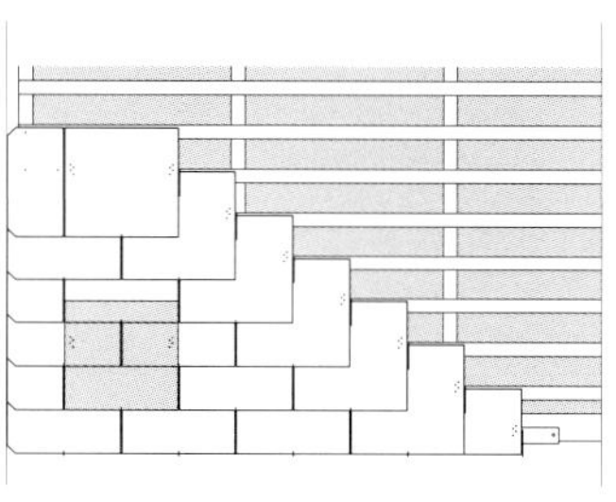

Double lap, Example format 40/40cm

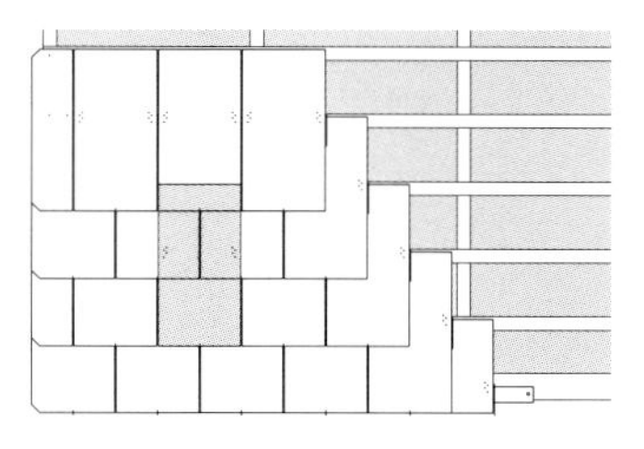

Double lap, Example format 32/60cm

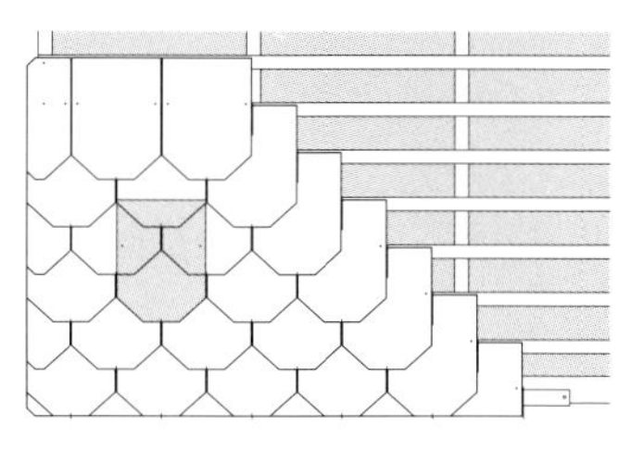

Double lap with truncated corners, Example format 32/45cm

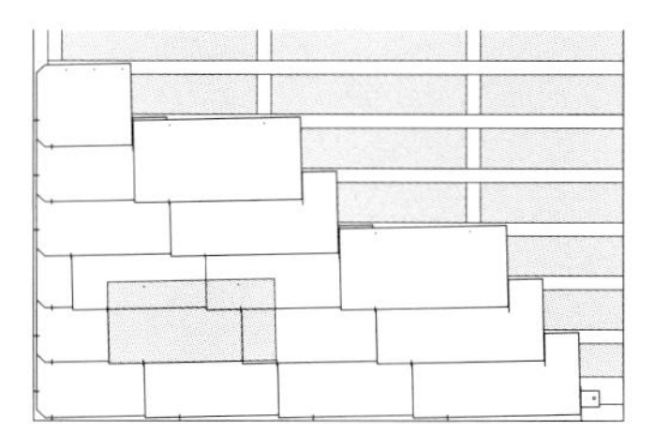

Horizontal lap, Example format 60/30cm

CORRUGATED SHEETS

Corrugated sheets are among the most popular fibre cement products all over the world. The traditional formats are

- Profile 5, max. format: 3100 x 920mm, thickness: 6.5mm, corrugation pitch: 177mm, corrugation depth: 51mm
- Profile 6, max. format: 2500 x 1097mm, thickness: 6.5mm, corrugation pitch: 177mm, corrugation depth: 51mm
- Profile 8, max. format: 2500 x 1000mm, thickness: 6mm, corrugation pitch: 130mm, corrugation depth: 30mm
- Short corrugated sheets, format: 625 x 920mm, thickness: 6.5mm, corrugation pitch: 177mm, corrugation depth: 51mm

The following loads can be used for design

- Corrugated sheets 0.20kN/m^2
- Short corrugated sheets 0.24kN/m^2

These loads are uniformly distributed loads, not including purlins, but including fastenings.

Head and side laps

The head lap for the corrugated sheets profile 5, profile 6 and profile 8 is generally 200mm. With short corrugated sheets the head lap is 125mm. For this head lap the corners of the corrugated sheets are given two cuts at the factory for left-hand roofing. Corrugated sheets and short corrugated sheets without corner cuts are available for roof edges. Any corner cuts on these corrugated sheets can be cut on site. The side lap of corrugated sheets profile 5, profile 6 and short corrugated sheets is always 47mm, which corresponds to about 1/4 of a corrugation. For corrugated sheet profile 8 the side lap is a full corrugation, which makes the side lap equal to 90mm.

Minimum roof slope

The minimum roof slope is the least possible roof slope at which practice has shown that a roof cladding would still exclude rain. The minimum roof slope for corrugated sheets is › 9°, for short corrugated sheets › 15° up to 10m rafter length. Additional measures to prevent rain entry are to be taken in the design and construction of clad roofs where increased requirements apply to the roof cladding. If suitable seals are incorporated in the head laps of the corrugated sheets, then the minimum roof slope can be reduced by 2° for corrugated sheets and by 5° for short corrugated sheets.

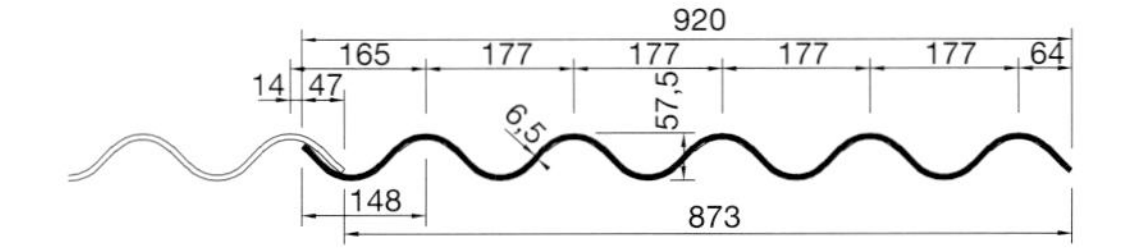

Profile 5 / short corrugated sheets

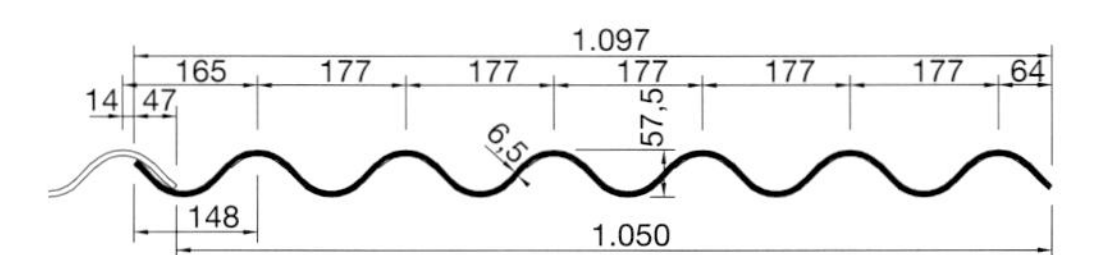

Profile 6

Support spacing
The effective length of the corrugated sheets is determinant in the calculation of the required support centres. The effective length is the actual length of the corrugated sheets minus the head lap. The roof slope also has an effect on the maximum support spacing allowed. The support spacing is always measured along the slope of the roof. Self-weight, snow and wind loads are also to be taken into account. With short corrugated sheets, the sheet length of 625mm and head lap of 125mm result in a consistent support spacing of 500mm, which ensures that short corrugated sheets can withstand the predicted wind and snow loads.

Ventilation openings
Roofs with corrugated sheet cladding can be constructed with or without ventilation to allow any thermal insulation system to function correctly. To ensure adequate and functional ventilation the openings at the top of the corrugations may not be fully or partially sealed, for example with a foam or mortar filler. The height of the open ventilation cross-section in the roof must be at least 2cm. The constriction caused by fitting ventilation grills, combs or other profiles in these openings should be taken into account particularly at the eaves and at the ridges of mono and double pitch roofs. The ventilation openings may need to be enlarged to comply with the minimum opening size requirements. Roofs with thermal insulation require an effective vapour barrier placed room-side. All service feeds and penetrations must be constructed as airtight. No condensation water must be allowed to build up as a result of diffusion or convection.

Fastenings
The type of fastenings used for corrugated sheets depends on the material of the substructure and the loads to be carried. For roof claddings with corrugated sheets the fastening is usually at the crests of the corrugations. Steel fastening components must be coated to protect against corrosion or be made of stainless steel. The fastening points are sealed with a plastic mushroom-shaped seal with a steel washer and cap. The distance of the fastenings to the edge of the corrugated sheet must be at least 50mm. Each corrugated sheet must have at least 4 fastening points. The fasteners must not be driven though the corrugated sheets. If corrosive gases, salts, other aggressive substances or persistent moisture is expected in the internal space under the corrugated sheets, e.g. in salt storage sheds, then stainless steel fastenings must be used.

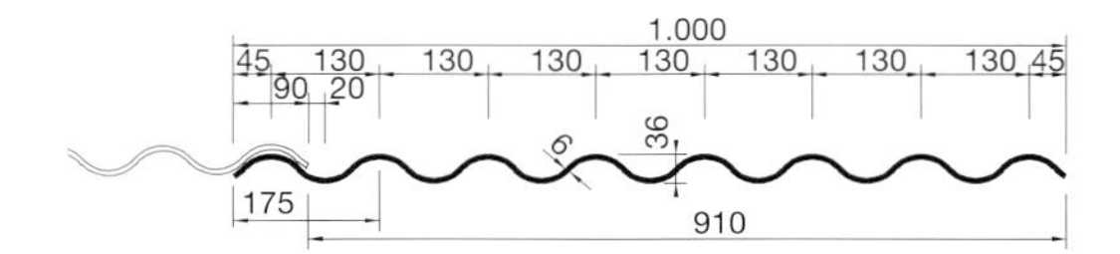

Profile 8

Galvanised hexagonal-headed woodscrews with a mushroom-shaped seal with a steel washer and small cap are generally used for fastening to wooden purlins. The diameter of the hole in the corrugated sheet must be 11mm. If insulation is installed on the cladding then longer fastenings are required. The distance of the fastenings to the edge of the corrugated sheet must be at least 50mm. The minimum cross-section of the wooden purlins must be 60 x 40mm. The depth of penetration of the woodscrews into the purlins must be › 36mm. If self-drilling woodscrews are used then the increased depth of penetration of › 57 mm means that the cross-section of the purlins must be at least 60 x 60mm.

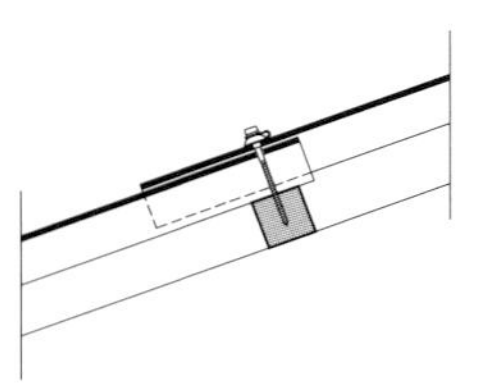

Corrugated sheet fastening on wooden purlins

If the truss spacings are long enough to require coupling purlins then the edge distances of the fastenings to the edge of the corrugated sheet must be at least 50mm. The requirements applicable to fastening the corrugated sheets to wooden purlins also apply to fastening them to coupling purlins.

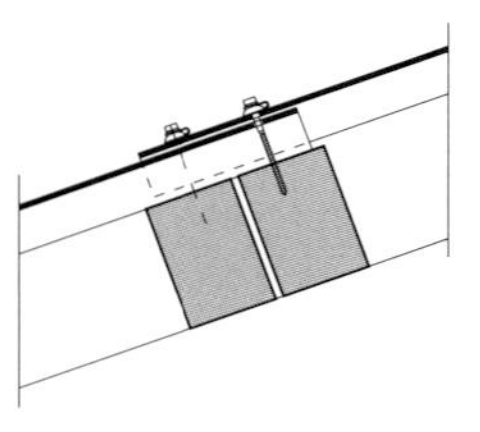

Corrugated sheet fastening on wooden coupling purlins

Fastening to steel purlins is normally done with steel hooks with a mushroom-shaped seal and steel washer. The following formula calculates the required length of steel hook: purlin height + purlin width + 22mm bending dimension + 90mm for corrugated sheets profile 177/51; purlin height + purlin width + 22mm bending dimension + 70mm for corrugated sheets profile 130/30. The predrilled holes in the corrugated sheets should generally be 11mm in diameter. Alternatively a self-drilling fastener with sealing washer, which has appropriate construction approval, or thread-grooving steel screws with mushroom-shaped seals and steel washers may be used.

Corrugated sheet fastening on steel purlins

Short corrugated sheets are fastened with 5 x 115mm coloured stainless steel dome-headed screws with sealing washers. Each short corrugated sheet is fastened through factory-drilled holes at the second and fifth corrugation crest.

Cutting and trimming

Corrugated sheets are available as standard with two factory-made corner cuts. These corner cuts are always for left-hand roofing. Left-hand roofing means that the corrugated sheets are placed starting from the right-hand verge and work proceeds towards the left-hand verge. Right-hand roofing means that the corrugated sheets are placed starting from the left-hand verge and work proceeds towards the right-hand verge. At the left-hand/right hand verge corrugated sheets without corner cuts are used as appropriate to the roofing direction. The required corner cuts are performed on site. The corner cuts are necessary to avoid four overlapping corrugated sheets at certain points and the undesirable loads this would create. Complete corrugated sheets, with their corners intact, are available for roof edge areas like eaves and verge. The roofing direction is determined separately for each side of the roof and is always given as the direction of working (left or right) from the verge towards the ridge.

Placing

The placing diagram shows an example of the arrangement of corrugated sheets profile 5 and profile 6 and for short corrugated sheets. If corrugated sheets profile 8 are used, then the placing arrangement can be derived from this. In this case, however, the side laps must be 90mm, a complete corrugation.

1. The first corrugated sheet is placed on the substructure.
2. The second corrugated sheet is placed, overlapping the first with a side lap of 47mm (for corrugated sheets profile 8 this is 90mm).
3. The third corrugated sheet is placed, overlapping the first with head lap of 200mm. The resulting joint between the second and third corrugated sheets should be 5-10mm.
4. The fourth corrugated sheet overlaps the second with a head lap of 200mm and overlaps the third by a side lap of 47mm. This arrangement allows the water to flow unhindered towards the eaves.

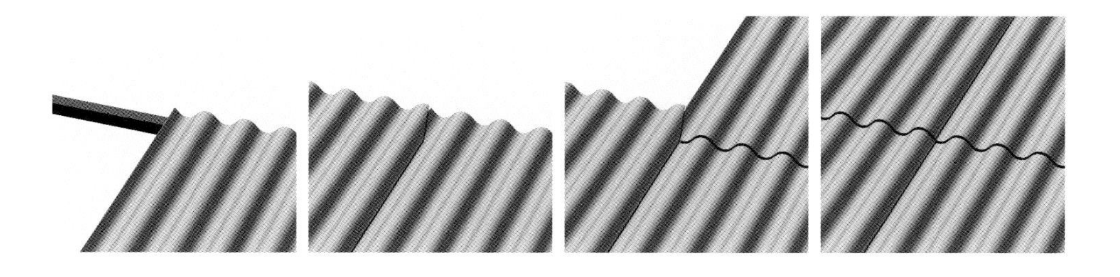

Placing sequence: 1-4

Eaves

The cladding at the eaves of a corrugated sheet roof can be completed with eaves closure pieces, eaves filler pieces, eaves ventilation combs or without preformed pieces. While eaves closure pieces and ventilation combs can be installed at the same time as the corrugated sheets, eaves filler pieces are installed afterwards. Whatever the detail adopted for the eaves, the roof must be provided with the required ventilation openings. The corrugated sheets in the starting course at the eaves must project beyond the roof construction so that water can flow freely into the roof gutters. When providing this outstand, the cantilevering length of the outstand must not exceed 1/4 of the maximum permissible support spacing for the sheets on that roof. According to this rule, the maximum outstand of corrugated sheets with a support spacing of 1150mm would be 280mm.

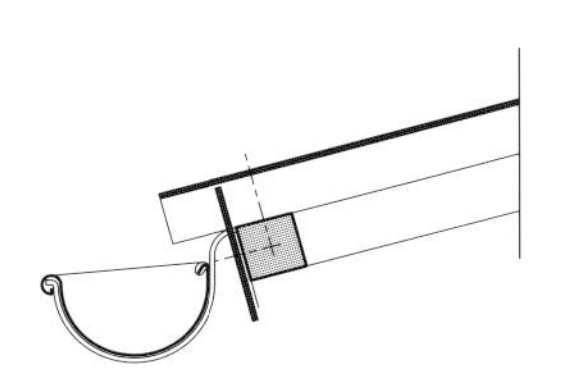

Eaves construction with eaves filler piece

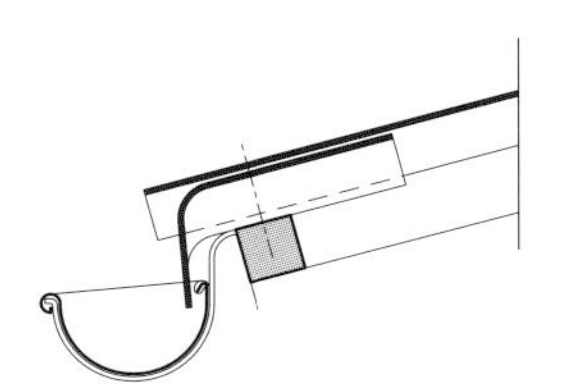

Eaves construction with eaves closure piece

Verge

The cladding of the verges of a corrugated sheet roof can be completed with gable flashing made from fibre cement. Alternatively the gable can be formed with projecting cladding or metal sheets. If the cladding is allowed to project at the verges, i.e. no preformed piece is used, then the roof cladding must end with a falling part of the corrugation. The last corrugation valley must always be supported on the substructure.

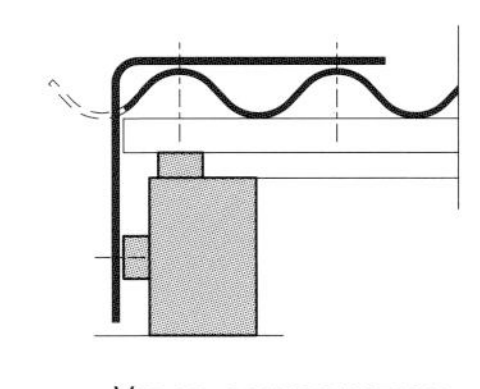

Verge construction with simple gable flashing

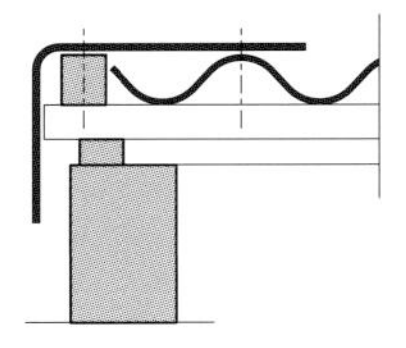

Verge construction with simple gable flashing and rising end corrugation

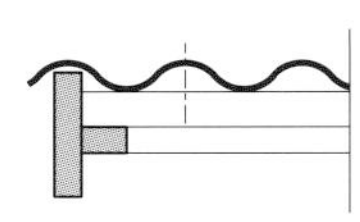

Verge construction with projecting sheet

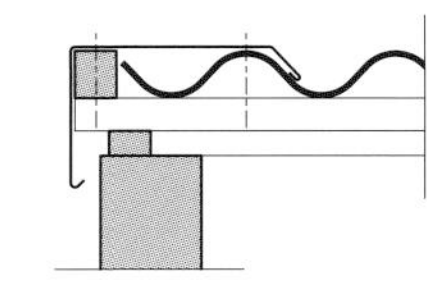

Verge construction with metal trim over sheet

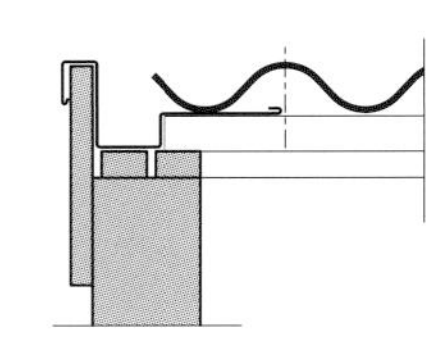

Verge construction with metal trim fixed under sheet

Ridge

The cladding of the ridge of a double pitch corrugated cladding roof can be completed with a 2-piece ridge with corrugated wings or as a cold roof ridge. Whatever the detail adopted for the ridge, the roof must be provided with the required ventilation openings. The 2-piece ridge with corrugated wings is independent of the opposite face of the roof. The 2-piece ridge with corrugated wings can be used with roof slopes of between 7° and 45°. Suitable ridge ventilation openings are available for all corrugated ridge pieces to achieve the required ventilation cross-section. A cold roof ridge can be used in the roof slope range of 7° to 40°. The ventilation cross-section is 250cm²/m per roof side.

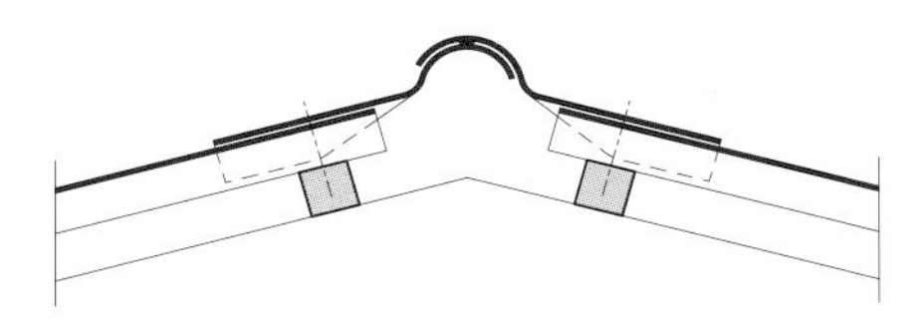

2-piece ridge with corrugated wings, profiles 5, 6, short corrugated sheet

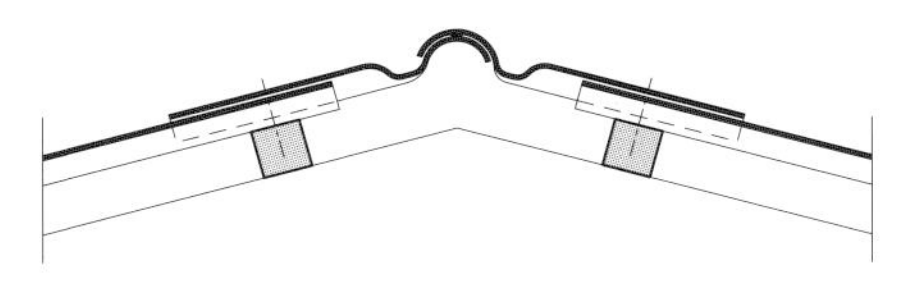

2-piece ridge with corrugated wings, profile 8

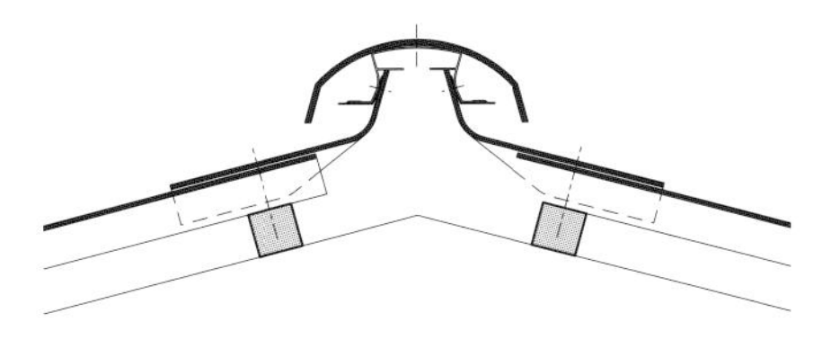

Cold roof ridge construction with corrugated sheets, profiles 5, 6

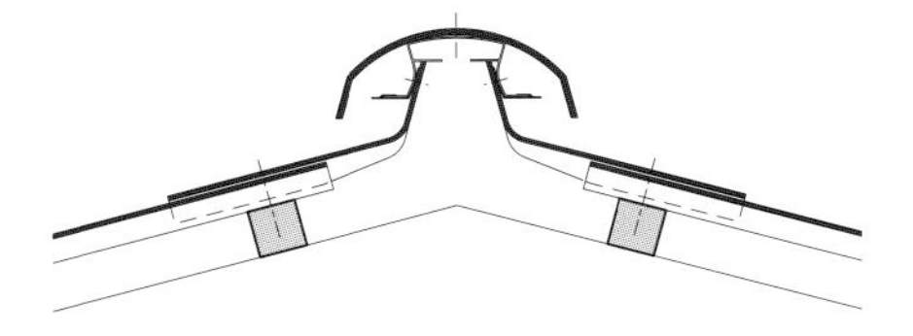

Cold roof ridge construction with corrugated sheets, profile 8

Preformed pieces

In addition to the above there are a great number of other specific pieces preformed out of corrugated sheets for various connection situations such as: corrugated apron flashing pieces, corrugated eaves bend sheets, corrugated window openings and bonded-on pipe supports, all made from fibre cement. Light-transmitting polyester and PVC corrugated sheets are available in the same profiles as the fibre cement corrugated sheets to allow light into the building.

CORRECT HANDLING OF ASBESTOS CEMENT

Before 1976 and the development of asbestos-free fibre cement technology, mineral asbestos fibres provided the tensile strength to fibre cement sheets. Until the 1980s asbestos cement played a leading role as a non-combustible construction material in buildings. Since the invention of asbestos cement in 1900 many millions of square metres of the material have been installed on roofs and facades. In Europe and in many other countries around the world the use of asbestos has been prohibited because of the hazardous effect on health which the dust generated during production and installation causes. In Germany, the guidelines and regulations for the refurbishment and maintenance of asbestos cement roofs and facades are some of the most extensive in the world and are considered exemplary.
In principle they say: there is no danger to health from installed asbestos cement products such as roof slates, corrugated sheets, facade sheets or pipes. Following risk assessments, the German Federal Ministry of Health has set a guide value of $1000F/m^3$ (fibres per cubic metre) for the environment for long-term exposure (24h/for life) that would not pose any danger to people. Contamination measurements near asbestos cement sheet applications have shown a value of $50\text{-}140F/m^3$. These values are around the detection limits for asbestos in the environment. This demonstrates that the weathering of asbestos cement presents no danger to residents or visitors.
The health risks that can arise from asbestos – asbestosis and certain forms of cancer – come from the fine asbestos dust that can result from improper treatment or handling. Fibrous fine dust is defined as having fibres of the following dimensions: length 0.005-0.3mm, diameter $<$ 0.003mm. The ratio of length to diameter must be greater than 3:1. The creation of fine dust must be avoided when carrying out refurbishment or maintenance work. Therefore the following rules must be observed if asbestos cement is to be handled correctly:

- High-pressure cleaning of asbestos cement products is prohibited
- Facades and roofs may only be cleaned with unpressurised equipment
- Uncoated fibre cement roof surfaces must not be coated
- Coated roof surfaces may be recoated if the substrate is suitable
- Coated facade surfaces may be recoated
- Uncoated asbestos cement products must be kept moist before removal. Coated products may be removed dry.
- Asbestos cement products taken out of a building must not be reused. They must be replaced with asbestos-free products. Asbestos cement products removed temporarily for maintenance purposes may be reused if they are undamaged.
- An approved and proven process must be used when drilling asbestos cement sheets, any other process that results in material being removed is prohibited.
- Breathing masks and protective clothing are to be worn
- The material must be packed and transported in dust-tight bags

Normally, a specialist company is engaged specifically to carry out the asbestos cement work in refurbishment projects of any significant size. The specialist company must be able to certify its work and have all the necessary equipment for dealing with asbestos cement available. Asbestos cement has to be stored under special conditions and be taken to a special disposal facility. Since its prohibition there are now industrial facilities for handling asbestos waste. These plants heat the asbestos in a tunnel furnace. The heat treatment sinters the material and completely destroys the asbestos fibres. The resulting product is harmless to health and can be reused, for example, in road-making.

Facades with fibre cement

The facade – our third skin – has a diverse role to play at the interface between inside and outside: it must protect from wind, rain, cold and heat, it should offer sound insulation and fire safety and beyond these building physics requirements, it should also tell us something: about the building itself, about its occupants or owners and about the times in which it was built. After the strict postulate of Modernism in which the facade had reflected the internal structure and function of a building, the facade today has once more been granted a high degree of autonomy. This new autonomy of the facade also brings with it a new autonomy of the material. Unlike many other materials fibre cement does not possess a definite meaning nor has it been saddled with one: it is neither urban nor rural, neither cheap nor extravagant. In this ambiguity lies a special opportunity to reinterpret the material in the context of the situation and design. The ability to cut large format sheets to any shape and fasten them in various ways opens up a wide field for the designer. Storey-high facade sheets are just as feasible as facade sheets cut into strips, free-form shapes small-format slates. Sheets can be mounted with rivets and screws, or attached from the back by undercut anchors, or glued in place with an adhesive system, the slates with slate nail and slate hook. Colourless or coloured varnishes applied to the sheets bring out the material's own character. Coloured cover-coats open the possibility of intensive colour designs with powerful colour accents. The selected examples show a wide range of different building projects and scales: from social housing in Holland to a prestigious corporate headquarter, from a shopping centre in Slovenia to a single family house in Hamburg.

Apartment block in Vienna, Austria

Design: Delugan Meissl Associated Architects, Vienna

The path to the realisation of this project was long. It was the third version of the design that was actually built on the Paltramplatz in the Viennese district of Favoriten. At first sight there is nothing to be seen in the solid corner block with anthracite-grey fibre cement facade of the original vision of an easily adaptable machine for living with a transparent glass facade. However, a second glance shows that the idea of flexible living still breathes through this design.

Projecting loggias alternate with recessed windows

The sculptural cube with cut-outs and embedded elements fits directly into the existing building lines of the block edge development. Unlike its historical neighbours however, it is composed of a summation of the elements: floor area, main shell and roof. The building springs directly from the ground unannounced by a plinth, and it is only prevented from reaching the sky by its overhanging roof. Framed by buildings from the post-1871 Gründerzeit, the first decades of the Prussian-German Empire, the building stands out because of its dark colour rather than its shape.

The relief of the facade puts life into the structure. Loggias pierce through its smooth surface sporadically. Their wide, dark aluminium frames break through the facade skin: a smooth surface of anthracite-coloured fibre cement sheets. In contrast to the loggia openings, the window apertures are set back in the facade and recede into the background. The entrance to the ground floor is recessed deep into the body of the building. These ground floor breaks in the facade are brightened in yellow and green. A ramp rises from the corner to accommodate the downward slope of the street. Colour guides the visitor on the ramp through a glass foyer and into the interior.

The sharp-edged cube contains 22 apartments. Flexibility, the central idea of the original designs, has been retained. Carefully positioned load-bearing walls allow the floor layout to be fully flexible. The wet areas are always arranged around the stairwell cores and belong to more than one residential unit. The 22 apartments differ from one another in their customised layouts from floor to floor. This diversity can be read from the facade. The loggia forms the connecting element between the interior living space and the outside world. The loggias are connected to the actual living areas by glass walls and are usually accessible from two rooms. Is it interior space? Or a balcony? Or just a terrace? The occupiers can decide for themselves. The glazing extends to the floor in contrast to the unusually high window parapets. The side walls and ceilings of the loggias are clad with brown-red wood, a material which creates an interesting tension with the dark fibre cement of the facade.

Right: Perforated flying roof over the black fibre cement cuboid

The load-bearing structure is reinforced concrete. When seen from a distance the monochrome grey walls have a concrete-like appearance. On closer observation however, the facade is revealed to be a skin of anthracite through-coloured fibre cement sheets. The architects were seeking to achieve a subtle contrast of surfaces. Instead of setting off the relief of the openings against colour or surface texture, the smooth panels provide a restrained background. The matt, shimmering lustre of the transparent varnished panels emphasises the reflections in the glazed openings of the loggias and windows even more clearly. Depending on the light conditions, the combination of anthracite sheets with the dark grey aluminium frames of the loggias produces various effects. Surface and structure are able to merge into one dark unit or enliven an interplay between foreground and background. At close quarters the surface resolves to a certain extent and the narrow joints reveal themselves between the variously sized sheets on which light and colour iridesce.

Composition of vertical and horizontal formats

Glass partition wall between living room and loggia

A careful combination of the grey fibre cement and other consciously selected materials emphasise the straightforward character of the design. The powerful effect of the loggias benefits from the surface having nothing applied to it. The facade sheets are fastened with rivets on to aluminium angle sections attached to the reinforced concrete structure. The placement pattern of the fibre cement sheets is just as sculptural as the building composition itself. Large format sheets of various shapes are sometimes arranged vertically, sometimes horizontally. The fibre cement sheets weave the facade image into a homogeneous whole and, with their subtle joint pattern, avoid any impression of monotony.
The architects have successfully performed the difficult balancing act between the individual demands

Material combination of fibre cement and glass

Open loggia

of the occupants and adopting a rigid layout. In the restrained design Delugan Meissl Associated Architects have found a formal expression for nuances. The resident's self-determination is promoted but not necessarily demanded. The building succeeds both in fulfilling the role of social housing and being a node in the wider social network. How intensively this web is spun from a resident's private space is individually controllable. A system of sliding doors allows playful interaction of room impressions and use. The roof terrace is the residents' communal space: sauna and relaxation areas are available to all residents.

PHOTOGRAPHS: MARGHERITA SPILUTTINI [2], DELUGAN MEISSL ASSOCIATED ARCHITECTS [5], VIENNA

SHADOWS AND REFLECTIONS

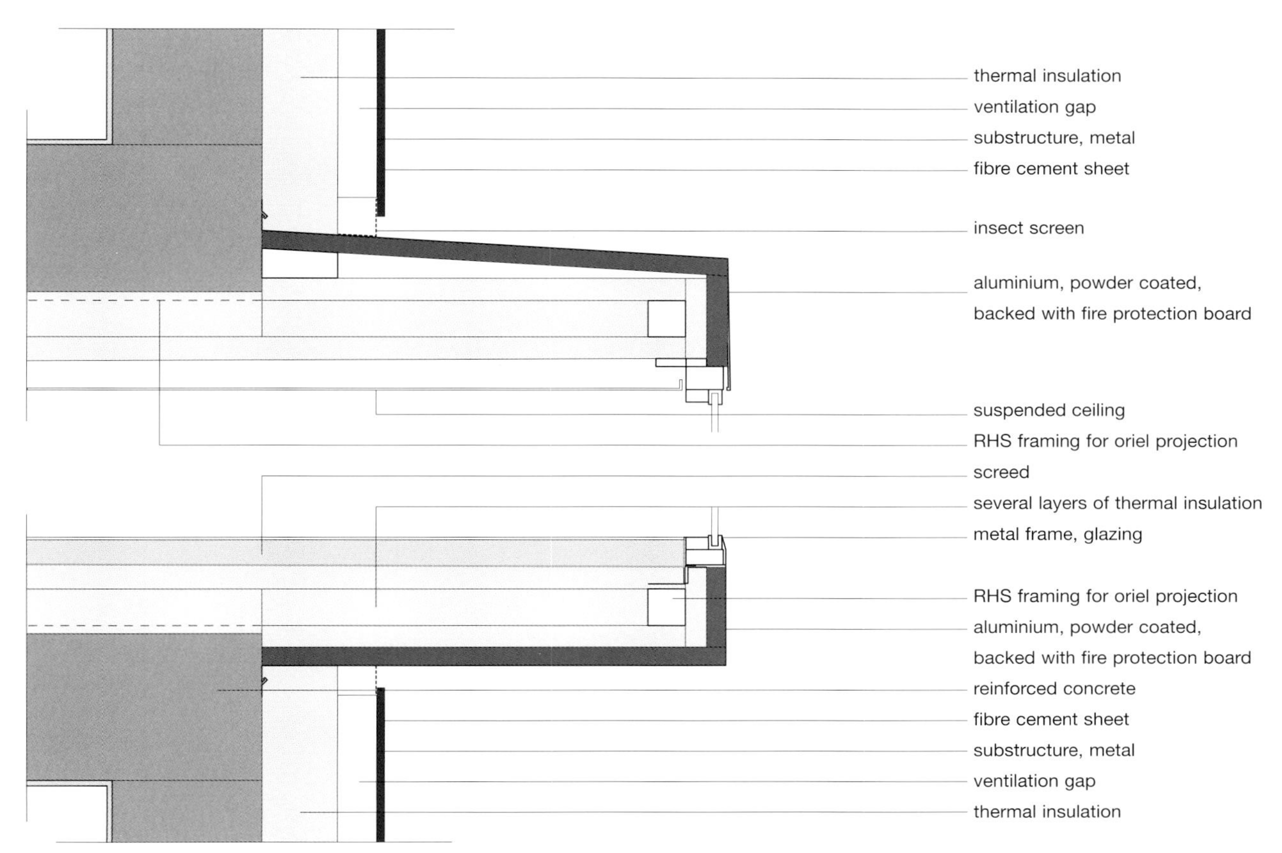

LOGGIA, VERTICAL SECTION
SCALE 1:10

FACADE, HORIZONTAL SECTION
SCALE 1:10

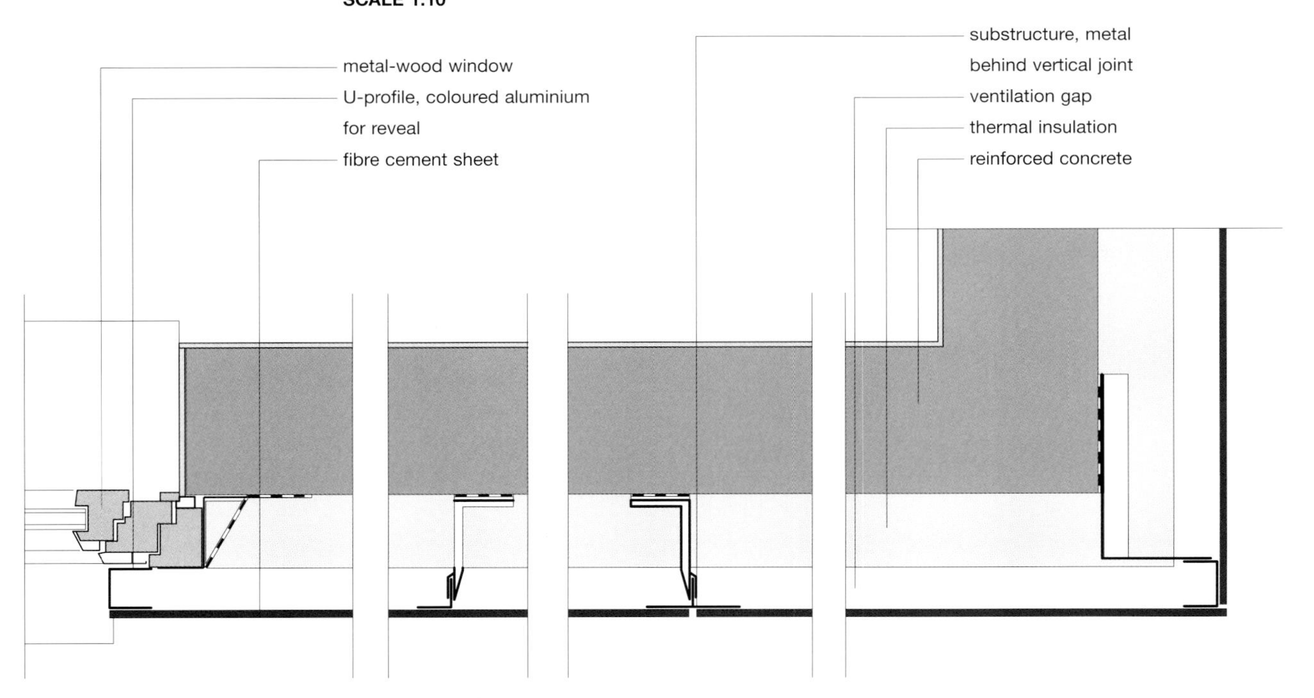

Residential high-rise in Amsterdam, The Netherlands

Design: Neutelings Riedijk Architecten, Rotterdam

High-flying sheets alternate with ribbon windows and recesses

Over the years, several cities in the Netherlands have seen the change of use and redevelopment of harbour areas release valuable inner-urban land. Amsterdam harbour, for example, was made available after its closure for redevelopment as a residential area. In the immediate vicinity of the Amsterdam Old Town district, about 8000 residential units providing homes for 17 000 people have been created on the former harbour islands and peninsulas of Java, KNSM, Rietlanden, Borneo and Sporenburg.

The IJ Tower by Rotterdam architects Neutelings Riedijk has become a landmark building, a widely visible symbol of this structural change. The residential building stands at the eastern end of the Oostelijke Handelskade on the longest of the harbour islands, directly in front of the KNSM island embankment. This site had long been identified as a focus for development in the urban master plan. For the time being the 70-metre high residential tower with its 20 storeys will remain the highest building on the peninsula. There

Right: Material combination: fibre cement sheets, aluminium strips, wooden loggias, stone plinths

are four apartments on each floor arranged around the internal access and services core of the 20 x 26m residential tower.

The building sends an impressive urban-planning signal with its height and facade design. Large format facade sheets made from fibre cement form the building skin. Dark-brown recesses and the horizontal pattern of the alternating bands of fenestration lend the tower its sculptural quality. The vertical grid of the fibre cement sheets offers an exciting contrast. The emphasis of the vertical joints creates the elegant impression of a pin-striped pattern.

As in the cartoonlike sketches with which the architects Neutelings Riedijk often approached their designs, the building has a strong figurative character. The facade, which appears two-dimensional on the drawing, also looks like patterned wallpaper in the finished project. The pattern of the alternating, horizontal ribbon windows on the vertically striped background wraps itself around the building. The windows sit flush in the facade and are normally divided into seven equal elements. The middle elements are blind windows, behind which partition walls between residential units or room-dividing walls can be connected. The bands of fenestration are carried around the corners, which gives the building a certain lightness, despite its volume. The corner is always formed with two returned windows. Large recesses into the body of the building break up this regular pattern. Shorter windows with only three elements connect to them, giving the simple arrangement a dynamic composition.

Pinstripe effect of vertical cover strips

The light-coloured, natural grey fibre cement sheets reach for the sky in narrow vertical strips, only 40cm wide. The ambitious character of the building is strongly emphasised by aluminium cover strips over the vertical joints. The patterns of the windows and the facade sheets are adjusted with respect to one another so that the windows always hit the middle of a fibre cement panel and not the joints between them. They look as if they have been cut into the outside skin.

The notches in the facade are up to ten storeys high and give the building an unmistakable appearance. The depth of these recesses is emphasised by a change to dark-red wooden facade sheets and can also be discerned from a distance. They have not only a graphic function but also play a real spatial role as the gen-

erator of a diversity of residential units that is scarcely evident from outside the uniform grid exterior. A total of 20 different floor layouts ranging from mini-apartments to luxurious maisonettes with roof terraces were generated for the 68 residential units. The arrangement of structural members in which, apart from the walls separating residential units, only one other wall per residential unit is load-bearing allows for later changes in the way rooms are divided.
The tower facade extends seamlessly from a three-storey attached building of glass and dark concrete masonry. With the tower proudly rising from the body of the attached building, the ensemble looks as if it were made in one mould. A maritime motif undergoes its urban transformation.
With the IJ Tower the architects Neutelings Riedijk have created a contemporary interpretation of the residential high-rise building. The sculptural form of the volume and the deceptive game played by the fibre cement facade are the external signs of an internal richness of variation unusual for this type of building. With their simple, well thought-through details the architects know how to counteract the usual image with unusual nuances.

Photographs: Christian Richters, Münster

Urban transformation of a maritime motif

Furniture store RS + Yellow, Münster, Germany

Design: Bolles+Wilson, Münster

It is a contradiction that furniture, which is above all an expression of taste and design, is often presented in banal commercial architecture. A surprising contrast is offered by the RS + Yellow furniture store in Grevener Straße in Münster, Germany, with its striking orange fibre cement sheet facade. The building itself suggests a piece of furniture in the city and in its form even makes reference to the separate sectors of the three retailers represented there and their various product ranges. With this design architects Peter Wilson and Julia Bolles-Wilson have produced an extraordinary ensemble, which gained an award for exemplary corporate architecture from the Federal State of North Rhine-Westphalia.

The great merit of the design is that in an environment which, despite its closeness to the inner city, already has the characteristics of the urban periphery, it reacts sensitively with its articulated form to its urban-planning surroundings. It defines a clear edge to the city, and inside the block the space used for customer car parking adds a market square element. A highly visible supersign, an advertising tower, opens a dialogue with the neighbouring angular tower of an 1839 clay brick gabled church, an old industrial chimney and a fire brigade tower. The main facade on the heavily trafficked road reveals the sculptural effect of the folds in the roof, store and tower.

The theatrical presence of the entrance is also intensified. Following a large roof hovering some 12m high, the passage between the furniture showrooms resembles a stage-like football goal area, whose conical narrowing creates a play with perspective that rejects any

Sculptural folding of roof, facade and tower

Right: Fibre cement as weather boarding, strips and large format ceiling in passage

Y
YELLOW

Material change in the inner courtyard

ideal viewpoint. Depending on the direction of viewing, either the street or the internal parking area is framed as an empty background. Window-like perforations in the orange roof similarly clad with fibre cement panels call the sky into the game of spatial reversals.

The three furniture stores have a total of 5000m² of sales floor and are spatially separated by the passage and parking piazza. With the design of the facade the architects not only set the urban accent, they make reference to the various types of furniture offered by the stores through their choice of material and facade details. This can be seen in the use of 35cm wide wooden frames for the showroom window of the natural wood furniture retailer RS. In this way Bolles and Wilson got around a local building regulation prohibiting the use of wooden facades on retail structures. The architects selected an appropriately

Fibre cement and ribbed glass plates

modern cladding of orange fibre cement panels for the furniture showroom "Yellow", which is well known for its youthful designs. On the other hand, the furniture retailer "Brands" was given a fibre cement facade of subtle anthracite through-coloured panels.
The powerful colour of the orange fibre cement sheets is based on a special surface coating technology developed by the manufacturer. The paint is applied at the factory in a multi-pass hot film, pure acrylate coating process, using exclusively UV-stable, coloured pigments. Mixing very fine glass spheres called Fillite, which is a lightweight filler additive, into the primer creates a slightly grainy surface with a water-repellent effect and little tendency to collect dirt. This breaks the surface tension of the rainwater on the facade. The rainwater forms droplets instead of flowing down the glassy surface in streaks and therefore the surface retains its brilliant colour.

The through-coloured fibre cement sheets on the other roadside facade are also given a factory-applied coating of pure acrylate. The matt transparent lacquer allows the authentic material characteristics of the fibre cement sheets to show through.
The architects chose various means of fastening the fibre cement sheets in place. Some of the sheets are flush butt-jointed, some fixed as weather boarding, casting distinctive shadows. The architects have covered large areas with narrow linear format sheets, which emphasise the length of the building. Larger format sheets are used in other places. The 8 to 12mm thick sheets are cut to shape in lengths up to 272cm at the factory. The aluminium substructure allows for different placement schemes. The horizontal aluminium profiles in the joints are coated black.
The large format sheets are fastened to the aluminium substructure using hidden fixings. They are suspended

Anthracite through-coloured weather boarding on the street facade

from the substructure by agraffes. The agraffes are attached by special undercut anchors, which are installed into blind holes predrilled in the factory. The same construction principle was adopted by the architects for the attachment of the fibre cement panels to the soffit of the large roof structure.

The overall design of the furniture store achieves its special quality from carefully designed details, in particular at the transition points between different materials. With a strong idea, an economic form of construction based on precast reinforced concrete units and an appropriate quality of execution the architects create a three-dimensional richness by their consistent interpretation and implementation of the retail shopping theme. The client also played a role in this successful part of the urban environment. For it certainly is not always the case that the client sees architecture as an integral part of retail culture.

Photographs: Christian Richters, Münster

Sculptural roof with window-like perforations

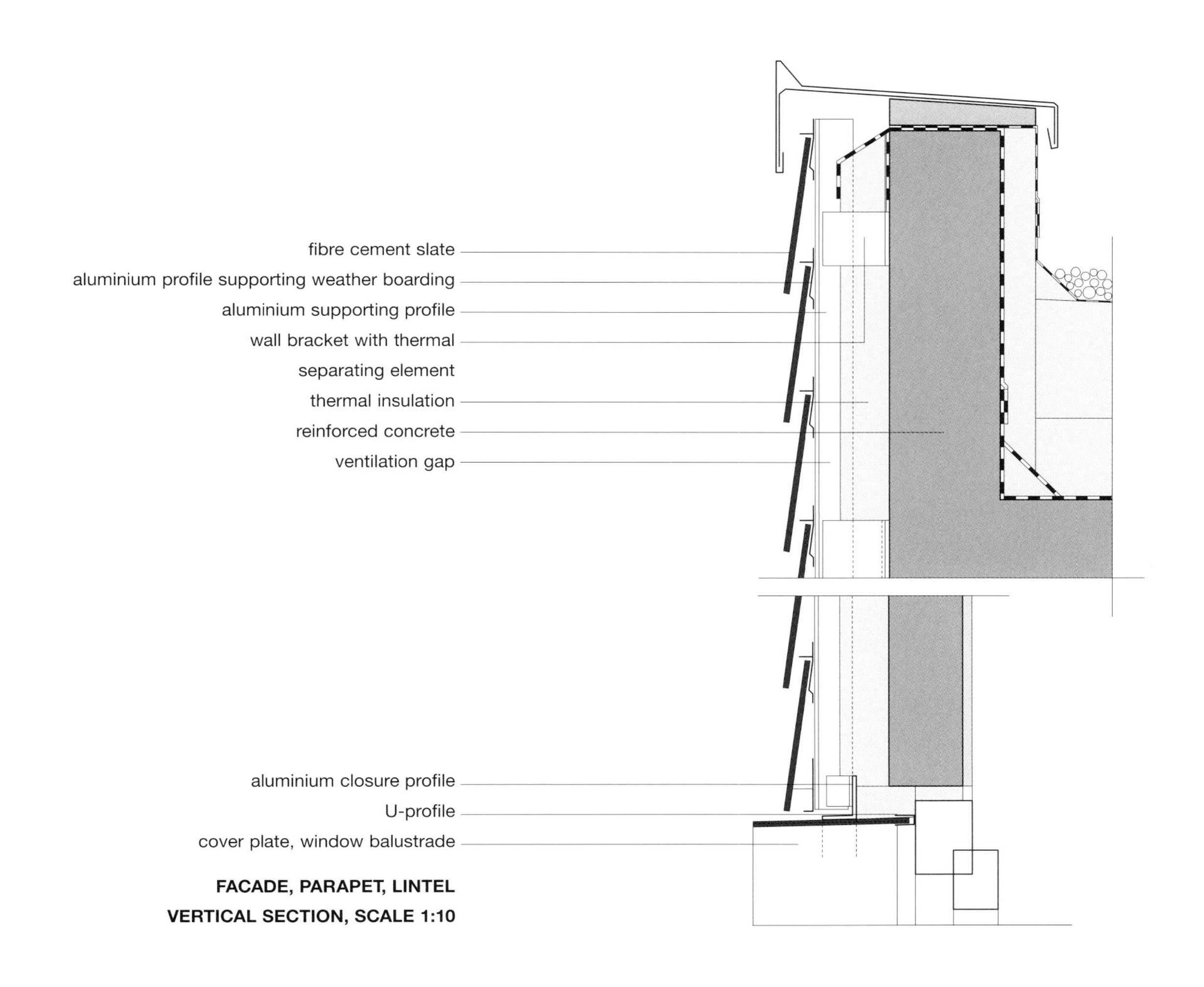

FACADE, PARAPET, LINTEL
VERTICAL SECTION, SCALE 1:10

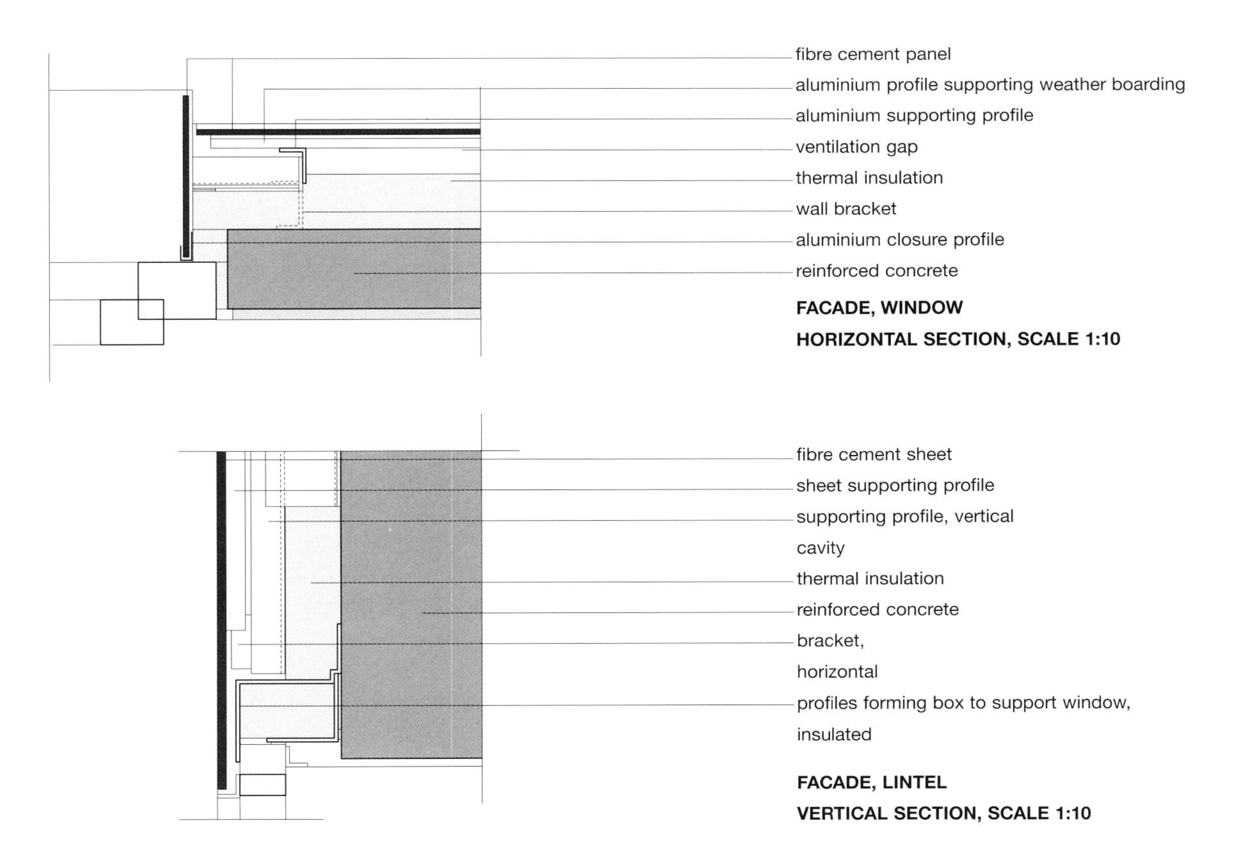

FACADE, WINDOW
HORIZONTAL SECTION, SCALE 1:10

FACADE, LINTEL
VERTICAL SECTION, SCALE 1:10

Caltrans office building, Los Angeles, CA, USA

Design: Thom Mayne / Morphosis, Santa Monica

The main offices of the Californian Ministry of Transport, Caltrans, takes up a whole city block in the centre of Los Angeles. The winning competition design accommodates just under 100 000m^2 of office space. Yet, this one-dimensional specification of use did not give rise to a uniform building. Instead, the lively structure of the city is reflected in the differentiated construction of the facades.

The building comprises a narrow ten-storey office block, which rises from a plinth and has four basement levels below. With an L-shaped floorplan, the building is organised internally into two main volumes. The larger part of the building extends over the length of the whole block and diverges at a slight angle laterally from the edge of the main street. Transversely to this, a four-storey block occupies half of the available forecourt. An open, elegant lobby cuts through the core of the building and links the forecourt with the main indoor entrance. At the sides of the open lobby, an art installation of horizontal, coloured neon tubes makes reference to the lights on the Californian freeways at night. A graphical element, a giant "100" sculpture rises up from the facade on the main street.

The flexible elements and irregular protuberances on the facade, which unfolds at pedestrian height into a protective roof, make the mobility of the city a theme with its lively structure. The staggered and fragmented surfaces are given a changeable appearance by the use of different materials.

Sculptural urban building block

Right: Plinth storey of anthracite through-coloured fibre cement sheets

1st St
FRIDAYS.
6219
Yellow
Cab Co.
808-1000

Horizontal band

Vertical plane

Varying textures of the facade materials contribute to the readability of the large building dimensions. Form, dynamics and surface define the building's functions and circulation routes, whilst the relatively uniform grey of the metal, fibre cement and glass surfaces creates visual cohesion.
Large format fibre cement sheets clad the three-storey plinth block. In its form of a sharp-edged box, it appears as a solid and substantial support to the main high-rise block. From the crossroad and the open forecourt the anthracite through-coloured fibre cement sheets form the perceptible, ground level surface of the external walls. The rectangular sheets were fixed in stretcher bond like stone tablets. They are fastened to a metal subconstruction as part of the cladding facade.
The seemingly closed surface is interrupted by long bands of fenestration. Attached flush with the plane of the fibre cement sheets, two perforated aluminium strips cover the glazed openings to further reinforce the appearance of the smooth, grey facade surface. The upper of the two strips of perforated metal is fixed in place whilst the other strip swings outwards pneumatically to allow an unobstructed view of the outside. The mobile aluminium facade reacts to the plinth and folds out over it as a roof. The south facade is coated with photovoltaic cells, which satisfy some of the building's energy requirements and screen the building from direct sunlight.
In particular, the diversity of materials and the interplay of the shapes of the building create a thematic reference to the Californian freeways. The building is extensively enclosed in a skin that looks homogeneous and yet is perceived to be in continuous change. Temperature and sunlight cause elements in the main building to open and give the building its chameleon-like appearance. Whilst at dusk the building skin becomes transparent and gives passers-by a hint of what goes on inside, the protective closed surface during the day gives no such insights. The facade creates an interplay with ever-changing views in harmony with the time of day and the weather conditions.

Photographs: Roland Halbe, Stuttgart

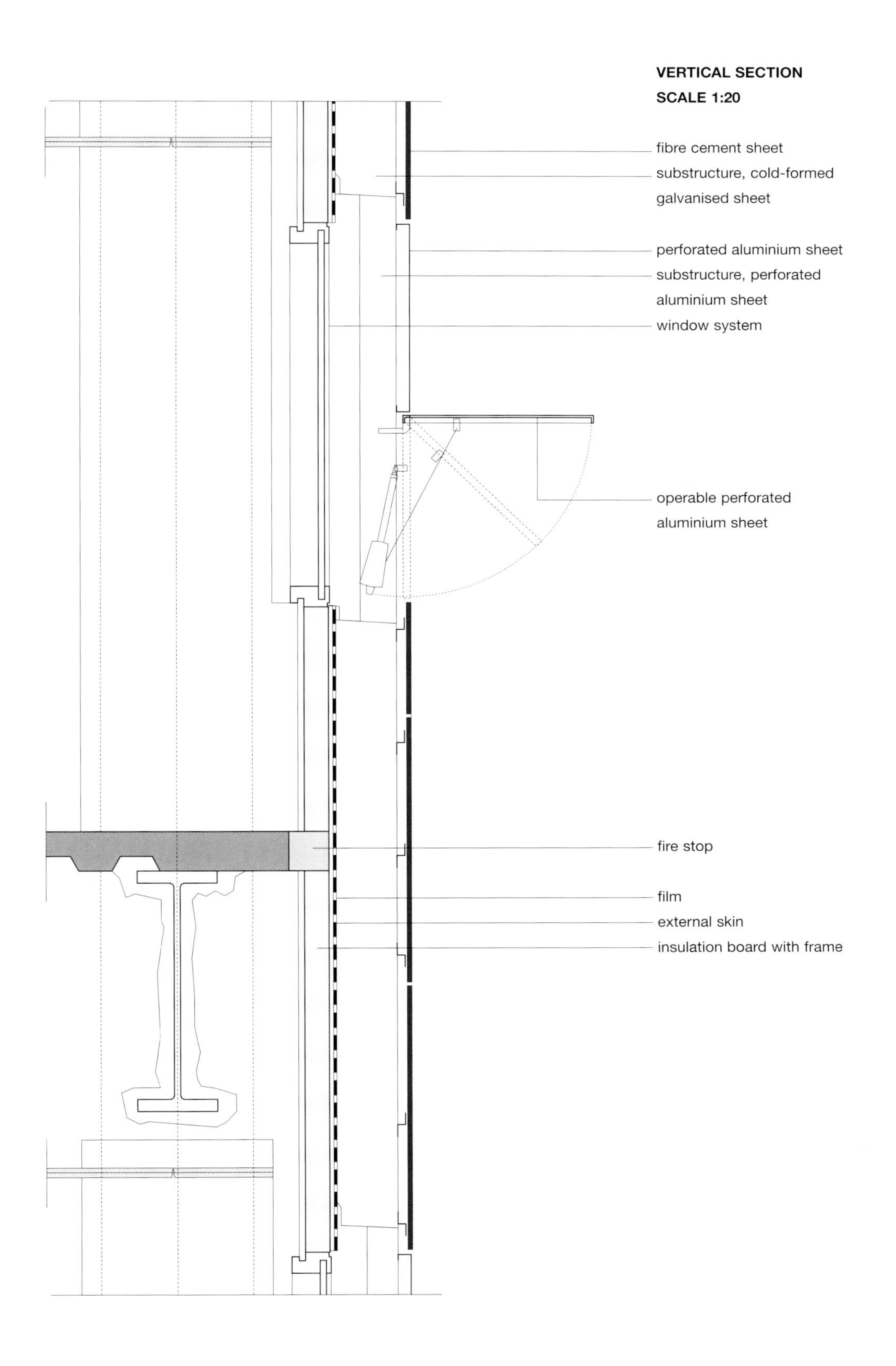
VERTICAL SECTION
SCALE 1:20
fibre cement sheet
substructure, cold-formed galvanised sheet
perforated aluminium sheet
substructure, perforated aluminium sheet
window system
operable perforated aluminium sheet
fire stop
film
external skin
insulation board with frame

Shopping centre, Nova Gorica, Slovenia

Design: Sadar Vuga Arhitekti, Ljubljana

Architects Sadar Vuga wished to create a building skin which was like a vacuum pack for the entrance building to the "Mercator Center" in Nova Gorica, Slovenia: indentations and bulges show where, inside the building, more space is needed and where less, where there is a lot of foot traffic and where not. The skin is thin and single layered – not covered with transparent film or flexible material but clad with fibre cement sheets which, with their red colour, exactly match the corporate identity of the operator.

From afar the striking building appears as a solid, red monolith, the body of which begins to resolve into individual details as the distance decreases. The fibre cement facade becomes fragmented, narrow entrances open up, intermediate spaces reveal themselves in the unfolded facade surfaces, through which natural light can enter the hall laterally. The entrance building creates a transit zone between arrival, parking and purchasing. The main building of the shopping centre is a completely refurbished precast unit warehouse from the 1970s. In multi-storey shopping centres, there can be an imbalance of shoppers frequenting individual storeys. So this entrance and circulation building, which also accommodates the technical services of the building, is sandwiched between the new multi-storey car park and the retail hall. Both parking decks

Closed impression from a distance

Right: Gill-like folds in the facade

cator Center

now offer direct access to the unconventional building which contains all the facilities for vertical circulation: escalators, stairs, passenger and goods lifts. In the entrance areas, where a great deal of customer traffic is expected, the floor extends over a greater area. This principle shows in the facade, which tightly encloses the underlying construction. The folded facade elements appear to enter into motion until they finally fold out completely at the entrances and transform into the canopy.
The interior of the building uses exposed circular ventilation ducts and unclad structural steelwork to create a highly technical appearance that matches its function. Expensive insulation is not necessary as the facade is single skinned and designed simply to be wind- and weatherproof, and must therefore – in contrast to conventional ventilated cladding facades constructed in fibre cement – be completely watertight. The constructional solution was specifically developed for this particular building. 15mm thick fibre cement sheets are attached to horizontal steel profiles and fitted with special joint cover profiles.
The regular panelled grid of the fibre cement facade clads the whole building, including the inclined surfaces. The 120m length of the building is divided into 38 approximately 3m wide strips. The intentional use of cover profiles at the horizontal joints provides additional emphasis of the horizontal aspect of the panels and the length of the elongated building. Only in the entrance areas, where the facade seems to fold out, does a vertical component come into play to create an exciting contrast. Here the panels are brought together in vertical bands which look as if they were made from folded paper. Daylight enters the interior of the building through the areas of plexiglass at the sides between the deployed facade elements and serves to highlight the entrance and exit areas. From the inside of the building the fibre cement sheets show their second colour, a natural grey.

Photographs: Miran Kambic, Radovljica

Smooth gable with open joints

Metal profiles in horizontal joints

GREY INTERNAL FACE OF THE RED FACADE SHEETS

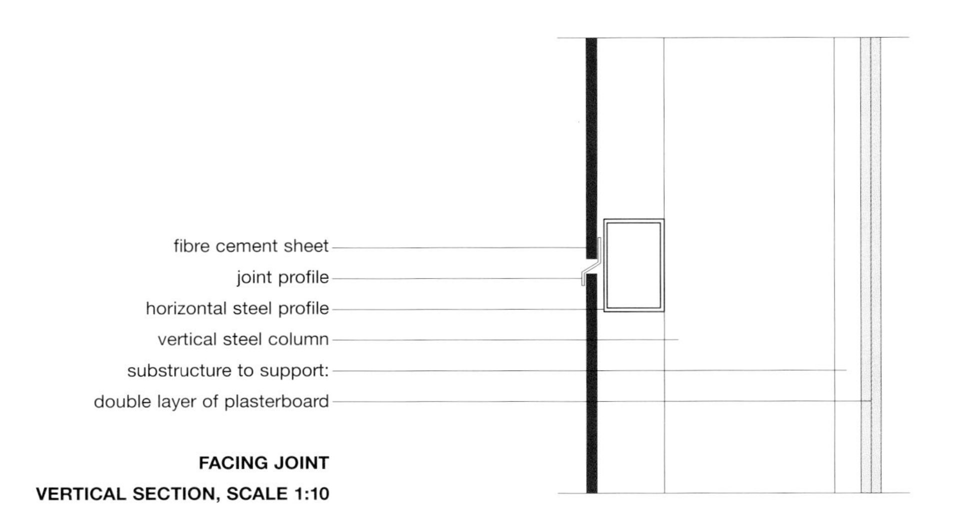

FACING JOINT

VERTICAL SECTION, SCALE 1:10

Neighbourhood centre, Stuttgart-Vaihingen, Germany

Design: Léon Wohlhage Wernik Architects, Berlin

To design a shopping mall along the lines of the US-American model was the brief of the design competition for the Schwaben-Galerie in Vaihingen: an introverted and fully air-conditioned block of around 23 500m^2 in area. Instead the Berlin architects Léon Wohlhage Wernik designed a suitably scaled and organised neighbourhood centre as an ensemble of several individual buildings, each with their own characters – a European version of the mall as an urban quarter, woven into its surroundings, with open walkways and piazzas. In a precisely crafted piece of urban design with a rich diversity of structure, the competition entry encompassed all the required functions and convinced the jury – with great benefit to the locality. The space was divided into five powerful areas: a shopping walkway with office space in the upper storeys, a hotel, the public civic forum, a market hall and a glazed atrium. For this part of the town with its characteristic pattern of small double pitch roofs, it was a huge leap of scale. However, the architecture emphasises building shape and lends the ensemble a distinct urban character.

The individuality of each building is highlighted by a sophisticated language of form and selection of materials. The one/two storey fibre cement facade cladding is the linking feature that brings the individual buildings into an ensemble. Through-coloured anthracite sheets or light beige varnished sheets, depending on the building type, characterise the plinth zones. Fibre cement defines the form and clads the whole facade of the civic forum: three sides of the building are clad to their full height with through-coloured anthracite fibre cement sheets. Here the material is consistently used right into the inclined surfaces of the shed roofs

Neighbourhood centre as an ensemble of individual buildings

Right: Civic forum with a facade of fibre cement, glass and exposed concrete

Urban space between market hall and civic forum

which allow light into the forum chamber below, whilst making reference to the past use of the site: the former bottling plant of a brewery once stood here.
Illuminated at night, the tall, glazed foyer of the 500-seat chamber is like a striking second skin in front of the anthracite-grey building. The dark, large format fibre cement sheets on the rest of the virtually fully clad building sides are arranged in a lively pattern and give the monolithic-looking building a special elegance in its detail. Whilst the horizontal joints run through at the same height, the vertical joints are staggered and fashion a facade rich in variation. Flush-fitted recessed windows do not disturb the planar surface of the fibre cement facade. Outward-opening tilting casements give the ventilated cladding facade a certain lightness. The projection room for the event hall is a particularly sculptural element where it breaks through the wall. The cement-grey exposed concrete box cantilevers at height into the street space and makes reference to the special function of the building.
The function of the market hall is also given particular emphasis by the shape of the facade. Profiled glass is a separate layer in front of a coloured rendered facade which sends a coded message in abstract letters into the city. Through-coloured anthracite fibre cement sheets in the plinth connect the building to the terrain. To follow the slope of the site, the fibre cement sheets are cut at an angle or stepped in the plinth area. The construction is ventilated and connected using visible rivets. If necessary, individual sheets can be easily replaced.
The shopping arcade and hotel plinths are less recognisable for the distinctiveness of their materiality. The colour of their rendered facades was chosen to match the beige of the fibre cement sheets. The difference is visible only in the joint pattern of the ventilated facade and the deeper window openings of the rendered facade. A special point to note is the stepless transition between the fibre cement facade and the external thermal insulation composite system. The change of material also coincides with the boundary between the areas frequented by the public and the less accessible areas. Following these design principles, the fibre cement facades are also taken a short way into the atrium. The external facade becomes an internal facade inside the atrium and reinforces the impression of the user being in an open square, albeit

one which is protected from the climate by a transparent glass skin. The ground floor has only narrow strips of fibre cement sheets between large showroom windows. In the storey above, the pattern is inverted, with wide strips of fibre cement sheets and narrow windows. Behind the beige-rendered perforated facade of the two upper storeys is generally office or hotel space.

An important criterion in the choice of a suspended ventilated curtain wall in fibre cement sheets was the ability to form flush joints with other materials. The cubic character of the individual buildings was reinforced by this approach. Another advantage was the minimal material thickness of the fibre cement sheets which, although full storey height in some places, was only 8mm. This is about the thickness of the windows and the profiled cast glass, which meant that the various layers of the ventilated construction could always be kept in the same plane. The thinness of the fibre cement facade gives the client a greater internal floor area, which is also to the benefit of the investors. Lastly, the through-coloured material has its own

Plinth of fibre cement as a linking element

Dynamic internal space

aesthetic dimension. Konrad Wohlhage speaks of an "optically perceived force that comes from within: through the depth of its surface, through its texture, through its velvetiness, which can quickly change with the light and weather".

PHOTOGRAPHS: CHRISTIAN RICHTERS, MÜNSTER

MATERIAL COMBINATION: RIBBED GLASS ABOVE FIBRE CEMENT PLINTH

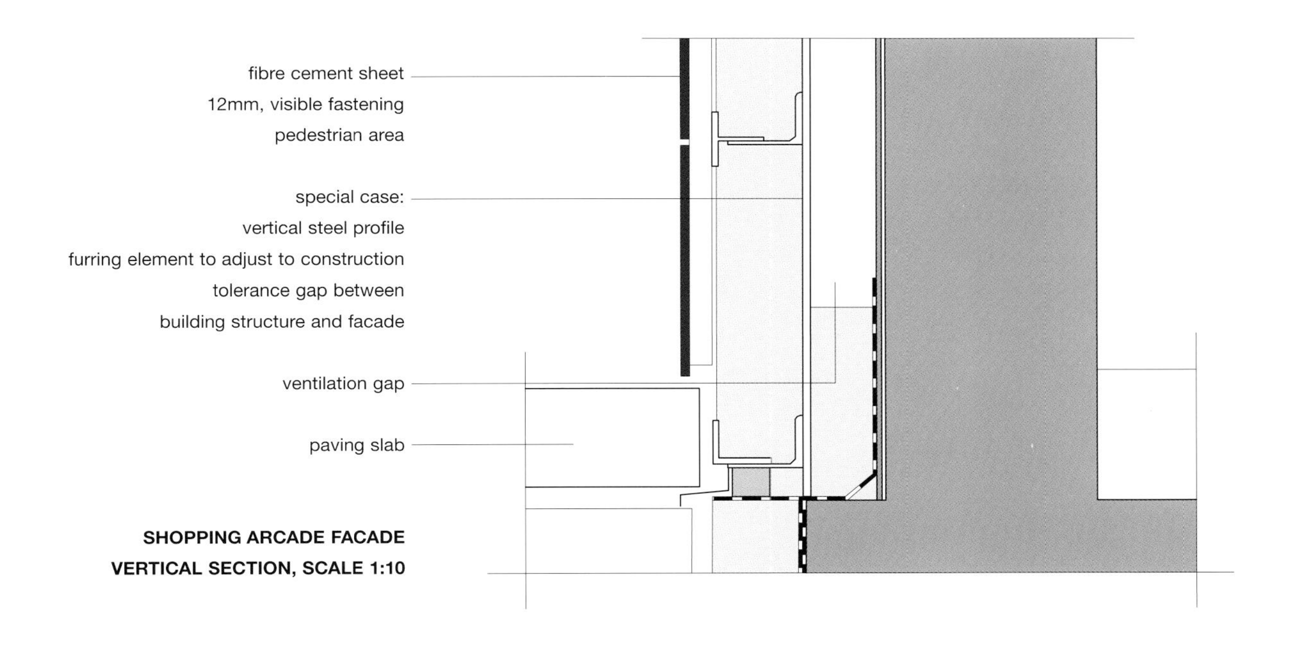

SHOPPING ARCADE FACADE
VERTICAL SECTION, SCALE 1:10

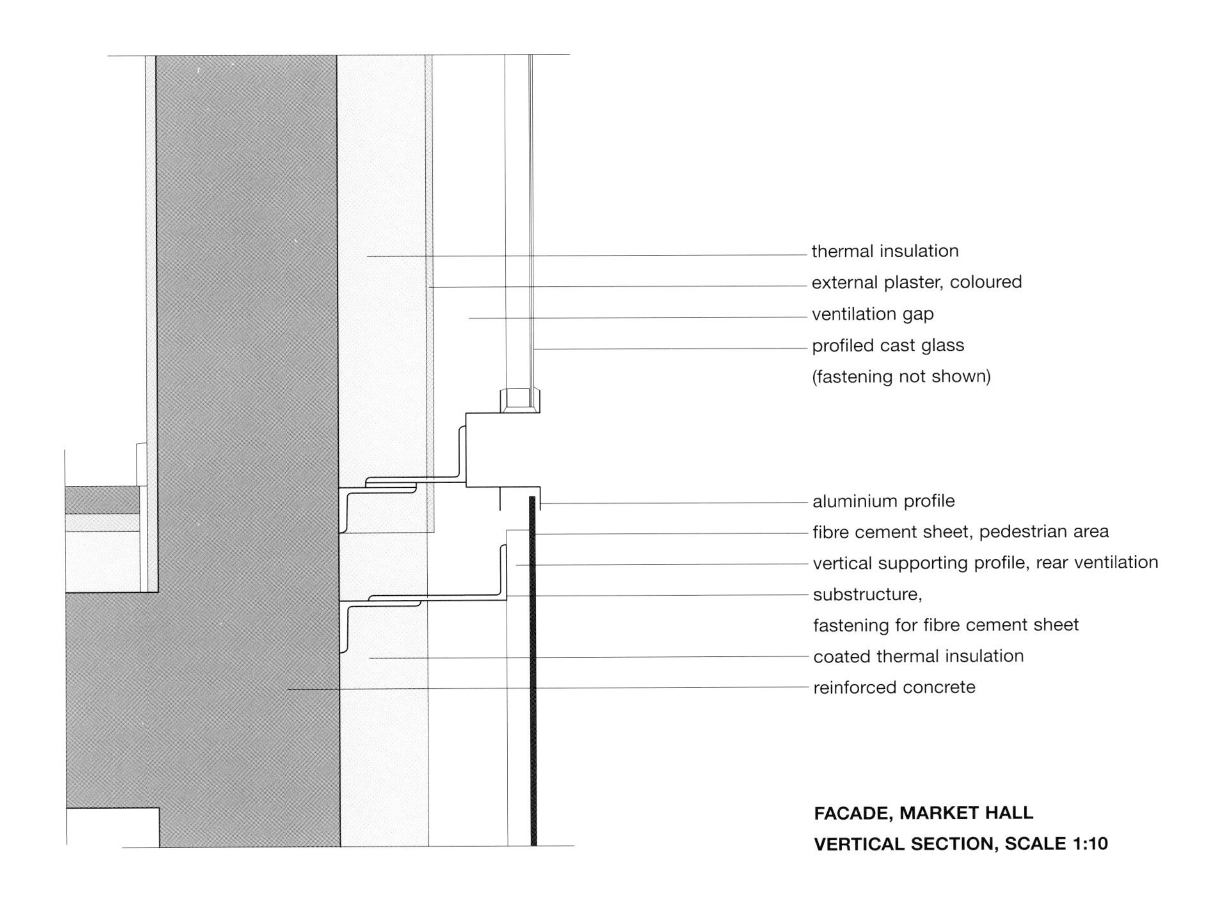

FACADE, MARKET HALL
VERTICAL SECTION, SCALE 1:10

Single-family house, Hamburg, Germany

Design: Kunst + Herbert Architekten, Hamburg

Composition of joints and formats with white cement sheets

One of the first buildings to have a facade of through-coloured white cement sheets can be found in the Hamburg residential suburb of Lemsahl-Mellingstedt. The district lies to the north of downtown Hamburg and is one of the Hanseatic city's attractive, middle-class residential areas. The house was designed by architects Bettina Kunst and Christian Herbert, who jointly run the "Büro für Forschung und Hausbau" practice in Hamburg.

The house appears genteel and reserved from the outside. An elongated structure, just one room wide and one storey high, it is topped by a slightly inclined roof and raises itself to a two-storey house. The garage is a cube on the entrance side. The top storey cuts across both parts of the structure and cantilevers over the entrance area. On the garden side, the architects meld these volumes, plastic on the north side, into a homogenous form. It creates a restful passe-partout for

Slender steel frames for window reveals

Precision in detail

the living room, which is glazed on three sides and which pushes itself towards the outside under a large roof terrace. Supported by circular concrete columns, the roof appears as an independent formal element, which exercises a sculptural power on the otherwise plain face of the garden. With the striking, translucent balustrades, the wide, metallically shimmering roof and the large glass surfaces of the extension, the house seems remarkably open and spacious. A large roof overhang ensures summer sun protection and creates a sheltered seating area on the garden terrace.

The timber planking on the terrace guides the visitor around the outside of the whole house, always following the white cement facade. Anyone who approaches the wall closely can appreciate not only the visual but also the haptic characteristics of the velvet-like fibre cement sheets. The ivory skin of the building seems to change its colour through various shades of light grey to snow-white, depending on the incident light and time of day.

The 8mm thick fibre cement sheets are riveted on to an aluminium substructure on the front face of solid limestone masonry walls. This form of facade construction is not widely used in private residential buildings but offers high thermal comfort in the living areas, is durable and requires little maintenance. In terms of building physics, the systematic separation of the structural walls, insulation and ventilated facade cladding is an effective means of preventing cold bridges and the formation of condensation water.

Whilst through their choice of matching colour aluminium rivets the architects have not developed the fibre cement sheet fastenings into a formal theme,

the dynamic composition of the joint pattern contributes to the lively character of the facade. The joints accommodate the lines of windows or spandrels but incorporate widespread staggers to create a diverse facade structure with sheet shapes and formats of various sizes and orientations. Like a mosaic, a lively and yet homogenous structure is created.
In a similar way to the facade sheets, the windows are of various sizes and orientation. In this case however, they follow a modular principle. A narrow vertical window extends to the floor and allows direct access to the garden from every room. Two or three such elements on the garden side and facing the terrace are combined into wider window openings. All sides of the window reveals are fabricated in the same metal frames, which maintain the planar effect of the curtain wall facade. To keep the clarity of the windows, the guardrail in the upper storey was made from 16mm laminated safety glass screwed on to U-sections in front of the window frame.
Deserving of special praise is the courage of the client, who, together with the architects, opted for a new approach in the rather conventional area of single-family houses. This was rewarded with an unmistakable and individually customised dwelling of particularly high quality.

Photographs: Oliver Heissner, Hamburg

Changing surface from ivory to snow-white

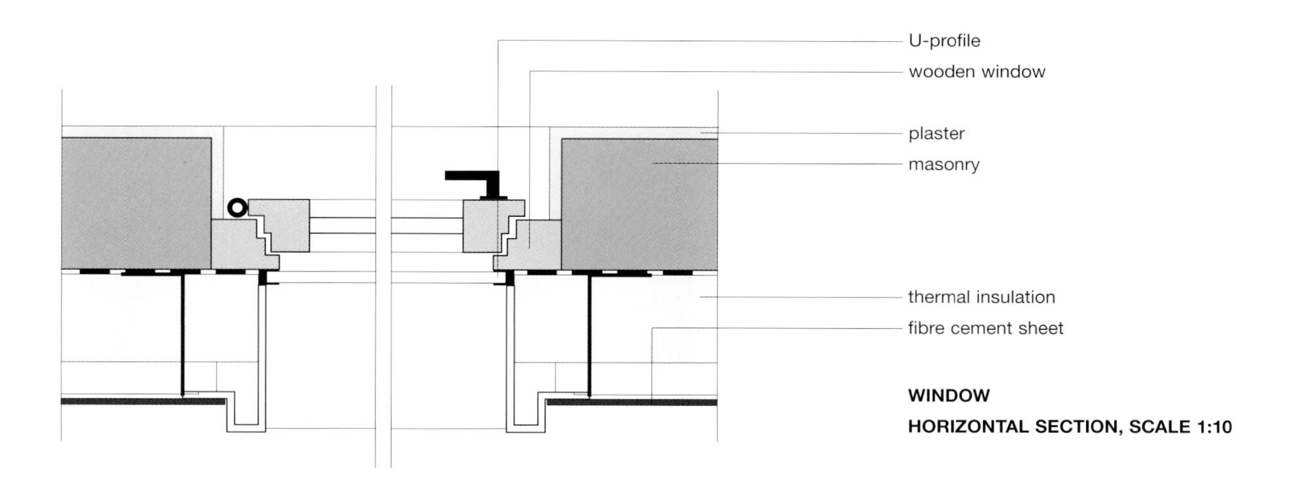

WINDOW
HORIZONTAL SECTION, SCALE 1:10

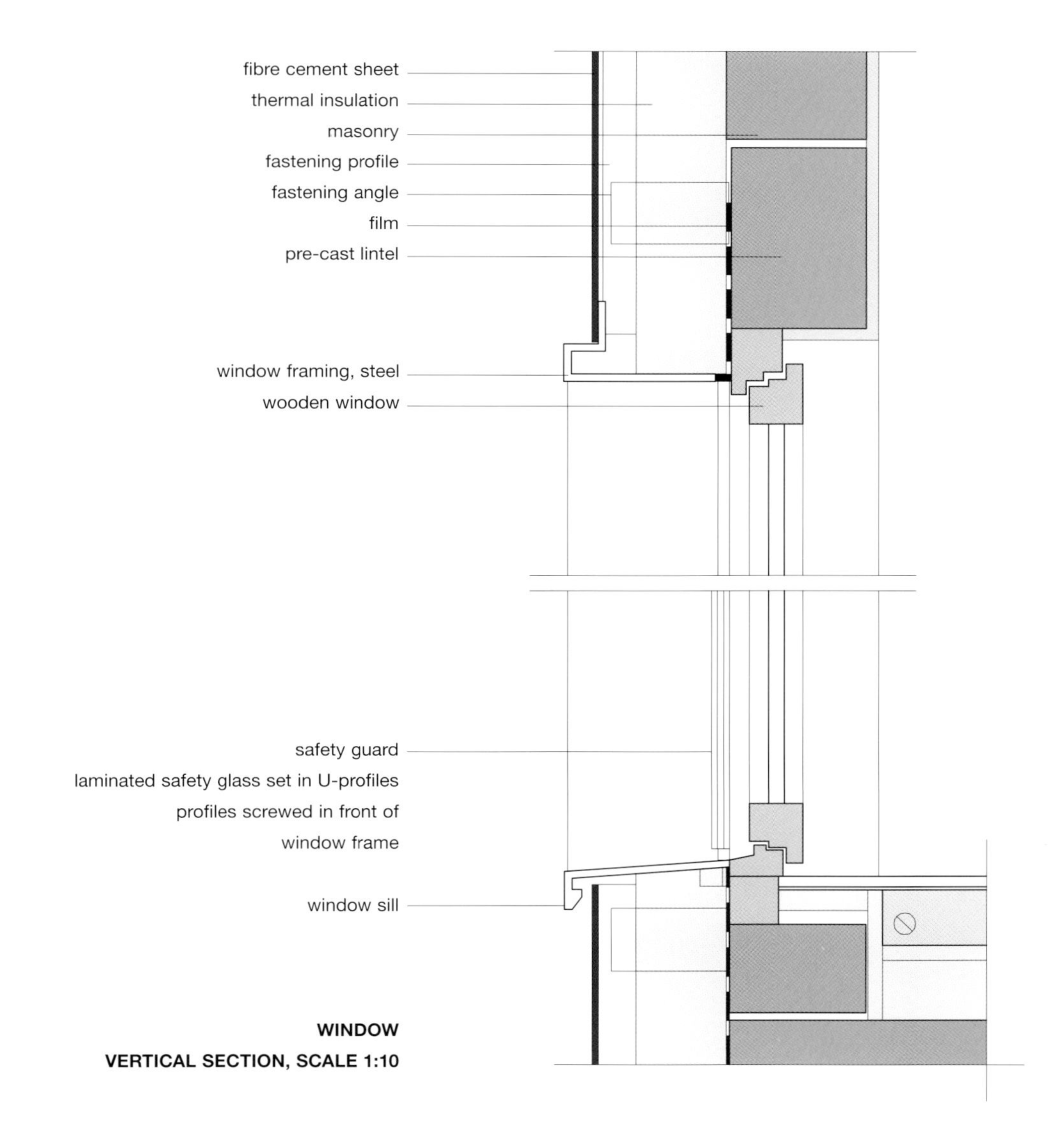

WINDOW
VERTICAL SECTION, SCALE 1:10

Vertical house, Venice, CA, USA

Design: Lorcan O'Herlihy, Culver City

Los Angeles is an extensive, horizontal city. Yet more and more people are returning to the city centre and building skywards. Dwellings closer to the city have many hidden qualities in comparison to the expensive and time-consuming commute from the far-off suburbs. "My home is part of an urban jigsaw puzzle," says architect O'Herlihy when describing the concept of his own house. To make the best use of the plot, the house butts right up to the boundaries and its roof terrace just manages to stay within the building height limit set by the local planning department.

Composition of open and closed surfaces

Right: Scale interrupted by slender, large format fibre cement sheets

On the narrow, somewhat constricted site, this house rejects the assumption that front and back must be different. The house looks like a black box from all sides – unusual for a dwelling in this neighbourhood. Dark fibre cement sheets characterise the facade. They were delivered in large sheets but were then cut along their lengths into narrow strips in order to accentuate the verticality of the building. White joints between the dark sheets run up the facade like the thin white lines of a pinstripe suit.

The closed skin of the "Vertical House" is perforated on all four sides by tall, narrow strip windows. Windows and fibre cement strips displace one another upwards and give the facade its characteristic pattern. The windows with their light-coloured, finely detailed metal frames continue from the delicate joints between the fibre cement sheets, beyond the intermediate ceiling slabs. They jump over the divisions between storeys, so that it is difficult to see how many floors the house actually has at first glance. Inside, these openings surprise the visitor as they emerge unannounced from the floor in one place and disappear into the ceiling in another. They each show a glimpse of the surroundings, like a work of art framed against the backdrop of a wall. In the views to the outside, the openings give the residents a feeling of openness, without threatening their private sphere. Light comes through the windows from the outside as if through a stencil. Some openings have translucent glass and others coloured.

The architect has made colour an "architectonic material". He uses mainly a dark coffee-brown in combination with yellow, blue and green accents in the facings. The sun shines on the coloured strips, which send their reflections into the house along ceilings, floors and walls.

In a playful rhythm, sheets and glass strips cover the skin of the building. The staggered openings demonstrate the clear separation between skin and structure. The load-bearing structure is a simple steel frame which forms the backbone for the plastered internal surfaces and provides the supporting framework for the external fibre cement sheets mounted on a plywood shell.

120

Interior effect of storey-spanning windows

Fine metal frames and joints

The differentiated use of a new material was the starting point for the architect when implementing the conceptual idea for the "Vertical House". He speaks of "manipulation of surfaces", as the architecture is determined through the skin and structure rather than through the actual building volume. The geometric, linear divisions are also found in the coffee-brown furnishings. Sliding doors and partition walls reflect the familiar divisions of the external facade. The kitchen ceiling lights are installed into recesses that imitate the format and arrangement of the facade sheets and openings.

Instead of a garden, the roof terrace reveals a view on to the Pacific Ocean, only three blocks away. Up there is a small extra room – like at the top of a lighthouse – a retreat looking out over the sea. The entire furnishings stand away from the outside walls so that open space is created without corridors. External and internal space are melded together by the meticulous spatial arrangement of the building elements.

Photographs: Michael Weschler, New York

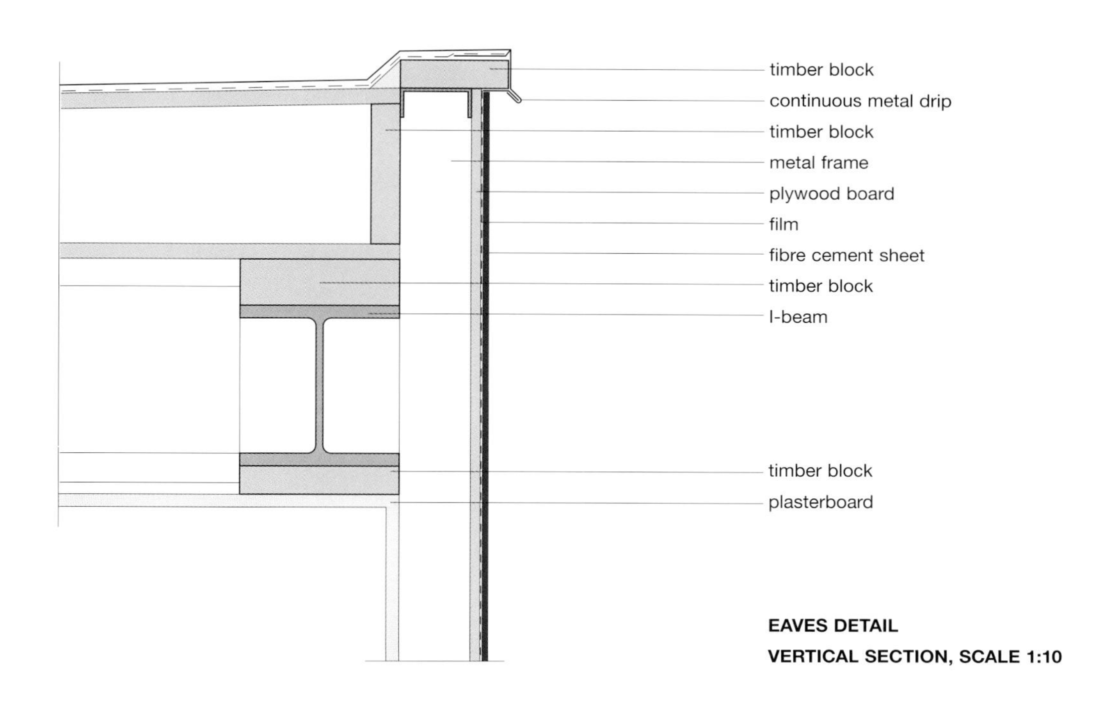

EAVES DETAIL
VERTICAL SECTION, SCALE 1:10

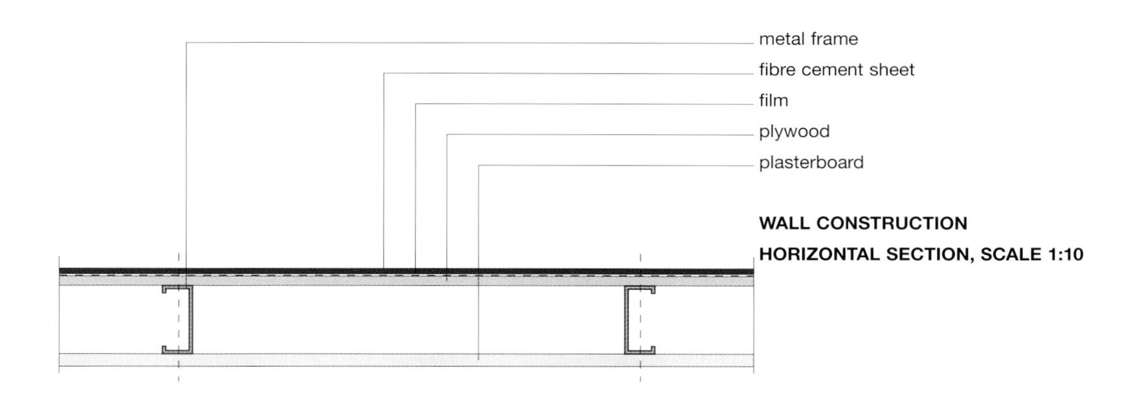

WALL CONSTRUCTION
HORIZONTAL SECTION, SCALE 1:10

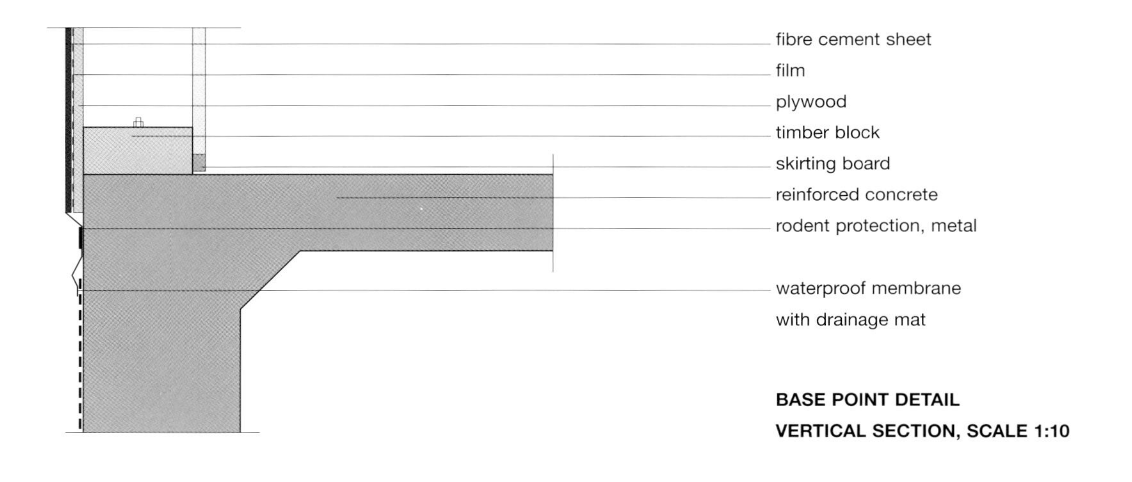

BASE POINT DETAIL
VERTICAL SECTION, SCALE 1:10

Residential estate in Viken, Sweden

Design: Tegnestuen Vandkunsten, Copenhagen

Stepped shades of grey: zinc grey, anthracite, black wood

How fibre cement can be used in a traditional context is shown by this example of a residential development in Viken near the Öresund Bridge in southwest Sweden, designed by the Copenhagen architectural cooperative Tegnestuen Vandkunsten. The former fishing village near the city of Helsingborg is characterised by narrow lanes and low, reed-roofed, half-timbered houses. A few steps from the old town lies a 7000m² parcel of land, the site of a former school, which is surrounded by old single-storey dwellings, trees and a local library. Here the architects proposed a row of double pitch roofed houses with black fibre cement facades, which would discreetly continue the small-scale features of the village.

The Danish architects became well known in the 1970s and 80s for residential construction projects with a pronounced social character. Instead of the social spaces typical of earlier projects, in the small town of

Material contrast: fibre cement sheets, glass, paving, wood

Viken they designed an intimate village community, which obtains its communicative character from its immediate neighbourhood and the resulting visual contact and over-the-garden-gate conversations.
58 residential units are organised in 25 houses of various sizes. The dwellings were constructed using prefabricated, locally assembled lightweight concrete and timber frame elements. Roofs and walls are highly insulated with fibres made from waste paper. From the outside the observer can scarcely distinguish between individual houses, only the variations in the roof slopes give an inkling of the different house designs. Terraced houses with low pitch roofs are mixed with semi-detached houses and apartments with roofs sloping at 45°. With its roof shapes and a conspicuously dense arrangement of houses with many nooks and crannies, the design makes reference to the traditional forms of buildings and development of this small town environment. On the other hand, material selection and detailing are thoroughly modern. Large glazed oriels at the corners of the buildings allow views deep into the interior. Zinc sheet roofs terminate almost flush with the facades. Anthracite through-coloured fibre cement sheets are the major influence on the appearance of the facades. The colour is a reserved, graduated range of shades of grey, which extends from the stone grey of the paving through the zinc-grey of the roofs and the anthracite of the external walls to the black striped garden furniture and timber carports. Only the oiled natural wood of the individually selected window casements set a slight accent of colour.
The colour and vitality of the fibre cement of the suspended ventilated facades are reminiscent of slate walls but have their own very special authenticity. The expression of the large format sheets changes depending on the moisture and the incident light on them. With no plinth, the facades extend almost to the ground. The leisure areas paved up to the house walls form a sort of public footway and private transition zone leading to the narrow asphalt-covered road surfaces. A horizontally oriented sheet arrangement was chosen for the consistently horizontal and staggered vertical joints in the planar facade grid. The width of

Building corners with glazed oriels

the sheets varies depending on the proportions of the wall surface and house type: a slimmer format than that of the long walls was adopted for the narrower gable walls. Windows and oriels look like recessed glass bodies.
On the inside, the 80 to 120m² units are efficiently laid out. The glazed oriels, storey-linking air spaces and room dividers that do not extend completely to the ceiling create a feeling of light and spaciousness.

In contrast to the floor layouts of the two-storey terraced and semi-detached houses, the floor layouts on both floors of the two-storey apartments are identical. The top floor under the roof can be accessed from external stairs. A patio replaces the garden, which is allocated for use by the ground floor apartment. The clear lines are continued into the external areas: orthogonally enclosed lawned areas use their geometry to break through the precision of the coarse gravel paving.

Photographs: Torben Petersen, Denmark

Fibre cement sheets at gables and long sides

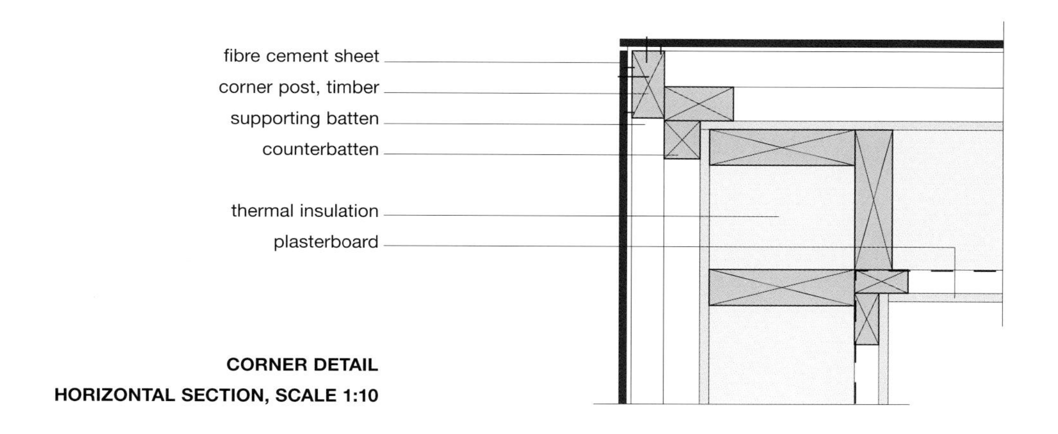

CORNER DETAIL
HORIZONTAL SECTION, SCALE 1:10

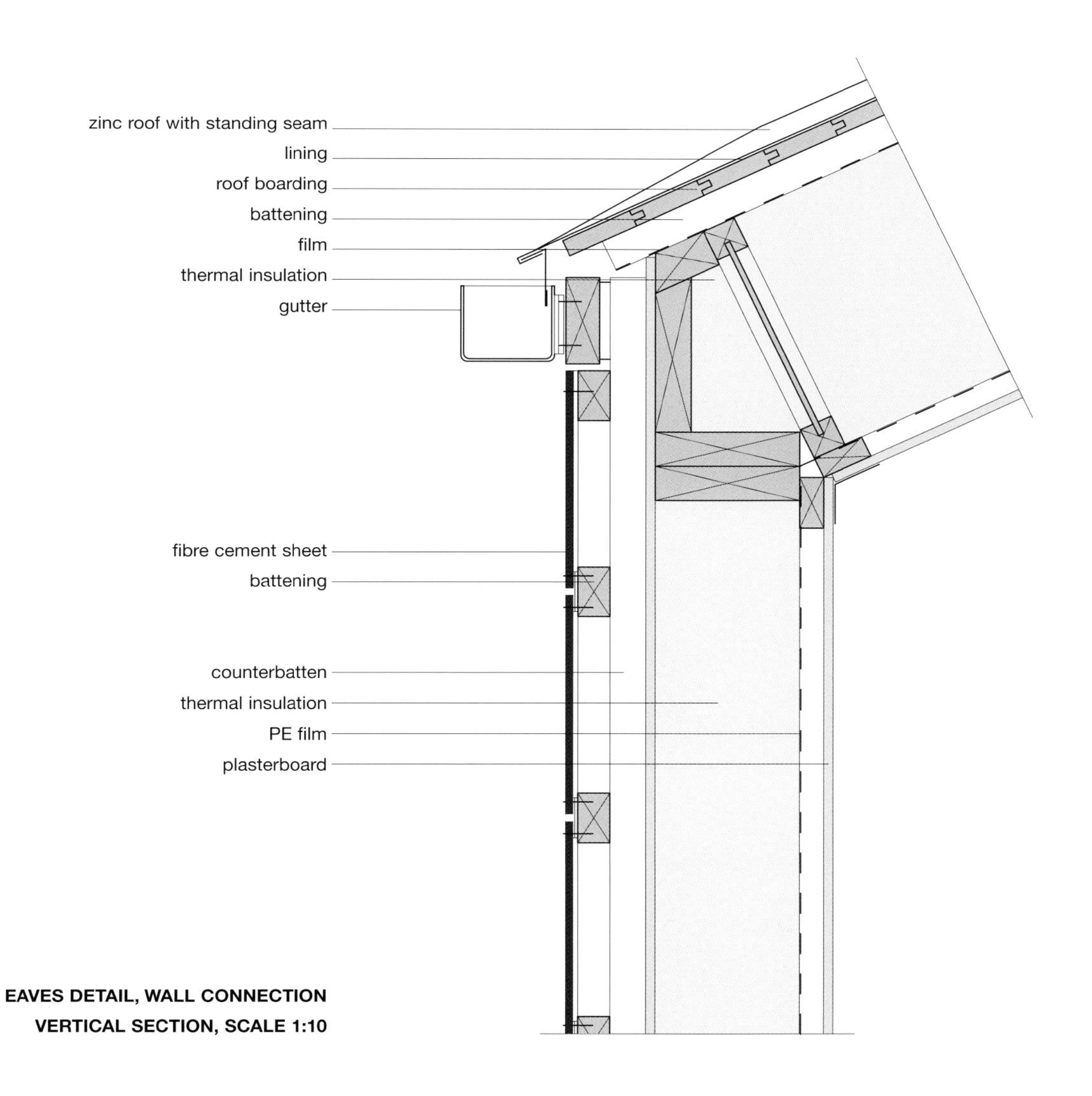

EAVES DETAIL, WALL CONNECTION
VERTICAL SECTION, SCALE 1:10

Community centre, Saint-Jean de la Ruelle, France

Design: Fabienne Bulle, Montrouge

Facade relief with closed sliding shutters of fibre cement

Architect Fabienne Bulle fulfilled the task of building a community centre in a socially deprived area with a mixture of pragmatism and poetry. The materials required to be durable, easy to maintain and inexpensive, as the available budget was extremely limited. Her feeling for rough, unfinished materials served her well. Above all with well thought-out details and subtle shades of openness and withdrawal, transparency and privacy, she has taken the habits and sensitivities of the users and inhabitants of the neighbourhood into consideration.

The design makes the transparency and unity of the facades a theme and interprets the building as a large piece of urban furniture, of which the outer fibre cement skin opens and closes by means of a sliding element made from the same material. This gives the community centre effective vandalism protection and allows different degrees of privacy and openness to be created for the building, which is mainly used as a youth centre.

A striking, voluminous, overhanging roof dominates the main facade. It fulfils two simultaneous functions

Open sliding shutters in front of the facade

Closed sliding shutters in front of the windows

by providing an inviting gesture to the city and solar protection for the large, south-facing windows of the general purpose room on the first floor. A narrow, lacquered band of steel projecting out of the facade and at the height of the storey ceiling leads around the corner to the side entrance, serving as a canopy and for lighting the entrances. The object-like character of the building is strengthened by the slim concrete slab on which it stands and the way it seems to hover a few centimetres above the ground. Like a black box, the house stands as a new urban building block in the district: independent, self-confident, without obvious reference to neighbouring residential buildings, and yet inviting.

All the closed facades are clad with fibre cement sheets. The architect composed the elevations as flat facade reliefs using through-coloured sheets and large glass panes. The red through-coloured sheets of the entrance facades stand in exciting contrast to the dominating anthracite of the rest of the building skin. The architect creates a lively, spatial interplay in the surface with sliding facade elements. The mobile fibre cement sheets of the red ground-floor facade are positioned in the front facade plane, flush with the fixed elements of the anthracite-grey facade. When they are slid in front of the windows the fixed red sheets behind them are brought into view. From a quick glance one would assume windows to be behind the red sheets. That the opposite is in fact the case makes this sliding picture puzzle so interesting. On the upper floor and at the rear of the building, the architect reverses the principle: here the red sliding doors run behind the plane of anthracite-grey facade and very effectively create the impression that the building, even with all the doors closed, does not look like a giant, windowless box but is rather more of an articulated house with openings and appropriate proportions.

The contrast between the open bays and the closed facade is particularly well brought out in the glazed corner on the ground floor. The changeable building continually adopts a new look in response to the time of day and current use. The materials were selected by the architect with durability in mind. "Fibre cement delights because of its special power of expression

Open building corner

and the authenticity of the raw material," explains Fabienne Bulle. The 8mm thick, door-high sheets are riveted on to an aluminium substructure. The sections supporting the ventilated cladding are screwed directly to the load-bearing concrete masonry walls. The regular joint pattern emphasises the clarity of the design. The electrically-operated sliding elements were developed by the architect.

Photographs: Philippe Ruault, Nantes

Closed sliding shutters

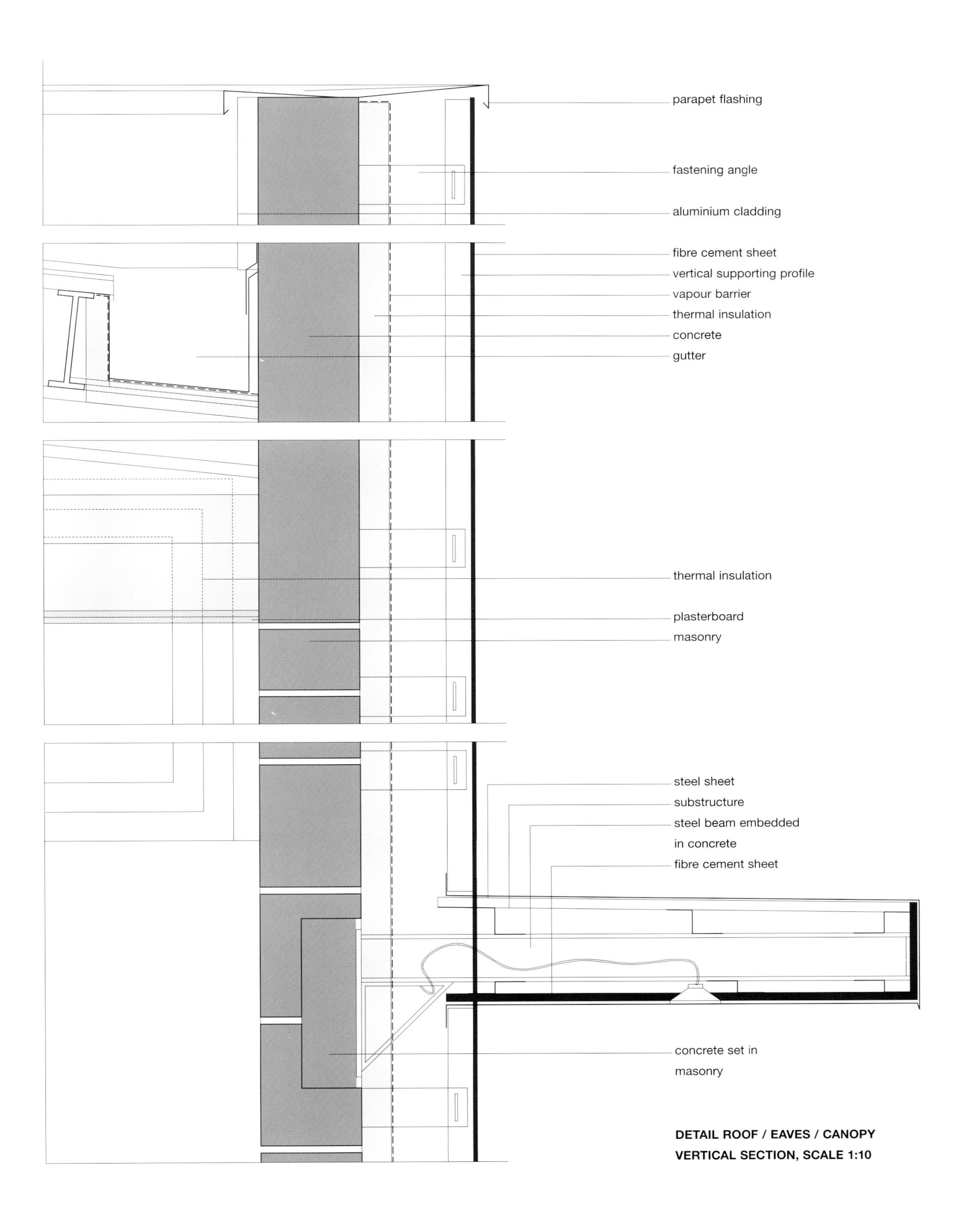

DETAIL ROOF / EAVES / CANOPY
VERTICAL SECTION, SCALE 1:10

Roofs with fibre cement

The pitched roof has been experiencing a renaissance for some time. Long since considered as un- or post-modern, it is being rediscovered as an archetypal form of a house, which has demanded some necessary reinterpretation. The roof is not only taking on a new significance in single family housing construction; even large buildings and whole urban areas are being defined by their roof forms. The return to the simple silhouette in no way means a simpler design. Refinement is hidden in the detail. Particular attention is being directed to the transition from facade to roof. Eaves and verge are not details requiring a purely technically solution; they are form-defining elements of the building and are of increasing importance to the whole design. Roof and facade may be treated as if they have been cast in a single mould. Only a few materials, like fibre cement, are suitable for both parts of a building. A skin made from a single material can have quite different appearances – depending on the choice of format. Today traditional roof slates are being almost simultaneously rediscovered in Germany, Belgium and Ireland as a scaly building skin. The corrugated sheet with its characteristic profile is making its entrance into the single family house scene as a textile-like curtain. Classical roof materials are being drawn into the facade. And in their turn large format fibre cement sheets are moving from the facade on to the roof. Thus the pitched roof is becoming a design element: the precisely formed fifth facade of the building. The selected examples demonstrate the contemporary relevance of the motif and the material, not only in single family houses but also on a larger scale in commercial buildings.

Single-family house, Malans, Switzerland

Design: Bearth & Deplazes, Chur

Corrugated sheet sculpture in an Alpine landscape

In the middle of the Swiss Alpine panorama an unusual red house reaches upward. Not only the colour but also the materiality of the corrugated fibre cement sheet facade distinguishes this building from the traditional architecture of the region. At the foot of the mountain sit the neighbouring houses, solid and conventional with wide eaves, some sumptuously decorated in the Engadine style. Facades in wood and in white render are the most noticeable features in view. A narrow path leads up the hill. Then the scene abruptly changes. Only at first glance does the shape devised by Bearth & Deplazes resemble the archetypical double pitch roofed house. Then the asymmetrical roof strikes the eye. It provides the first clear hint of further characteristics departing from the norm.

Sharp contours define the house from the base to the roof. The red corrugated sheet facade forms a sealed shell. Only at the deeply recessed areas of the entrance, garage and loggia does the outward-facing, white interior offer light-coloured, planar surfaces as a transition from outside space to inside space. The luminous red colour of the corrugated sheets does not have a disturbing effect on the neighbourhood, rather it links the complementary green of the vineyard behind the house with the sunflower field in front. Only the faint shadow gaps between the corrugated sheets are visible. The crests and valleys of the corrugated fibre cement sheets fit exactly into one another, making vertical sheet transitions scarcely visible. The corrugated sheets wrap around the house like horizontal

Right: Red corrugated sheets in horizontal bands

Sharp lines at the eaves and verge

bands. At the corners of the house the ends of the corrugations overlap to close the skin without a joint. The sheets are screwed on to a wooden substructure. The repetition of the rising corrugations sets its orderly pattern against the slightly misshapen house.
The steep double pitch roofed house appears to be an independent part of the overall shape of the all-embracing building skin. At the eaves and verges, the profiled sheets simply show up as a narrow, knife-edged line on the facade. Again and again the line of the eaves leaves its horizontal track on the asymmetrically shaped building. On the east side, the eaves are interrupted by a high dormer window that rises straight up out of the wall. Opposite, where the south gable and west facade meet, the eaves rise diagonally to make space for the loggia below. Roof gutters and downpipes are hidden behind the ventilated cladding facade. This precision in the details contributes to the homogeneous appearance of the house.
In order to enclose the shape like a seamless layer of clothing, the 2.5m long corrugated sheets were selected to be as large as possible. The cladding accommodates the irregular shapes caused by the external wall departing from the rectangular and can be aligned to follow exactly all the inclined edges and kinks. Using no standard preshaped pieces, the corrugated sheets are cut at an angle and joined directly to the edges of the building. The windows with their thin black profiles look as if they have been cut into the facade skin with a scalpel. They break away from the preset pattern of the corrugated sheets. The fibre cement panels are cut precisely at an angle and fixed around the windows.
Although the material character of the fibre cement skin behind its intensive colour coating is not always apparent, the house has a look suggestive of solid masonry. Unlike some corrugated metal facades this one does not give the impression of being temporary. The shadows cast by the corrugated sheets lend the skin a textile lightness. On the other hand, the shape and stone materiality of the house confirm its place as part of the rocky mountain landscape. In this project, the delicate yet robust fibre cement material embodies the relationship between building shape and skin, between substance and material.

Photographs: Ralph Feiner, Malans

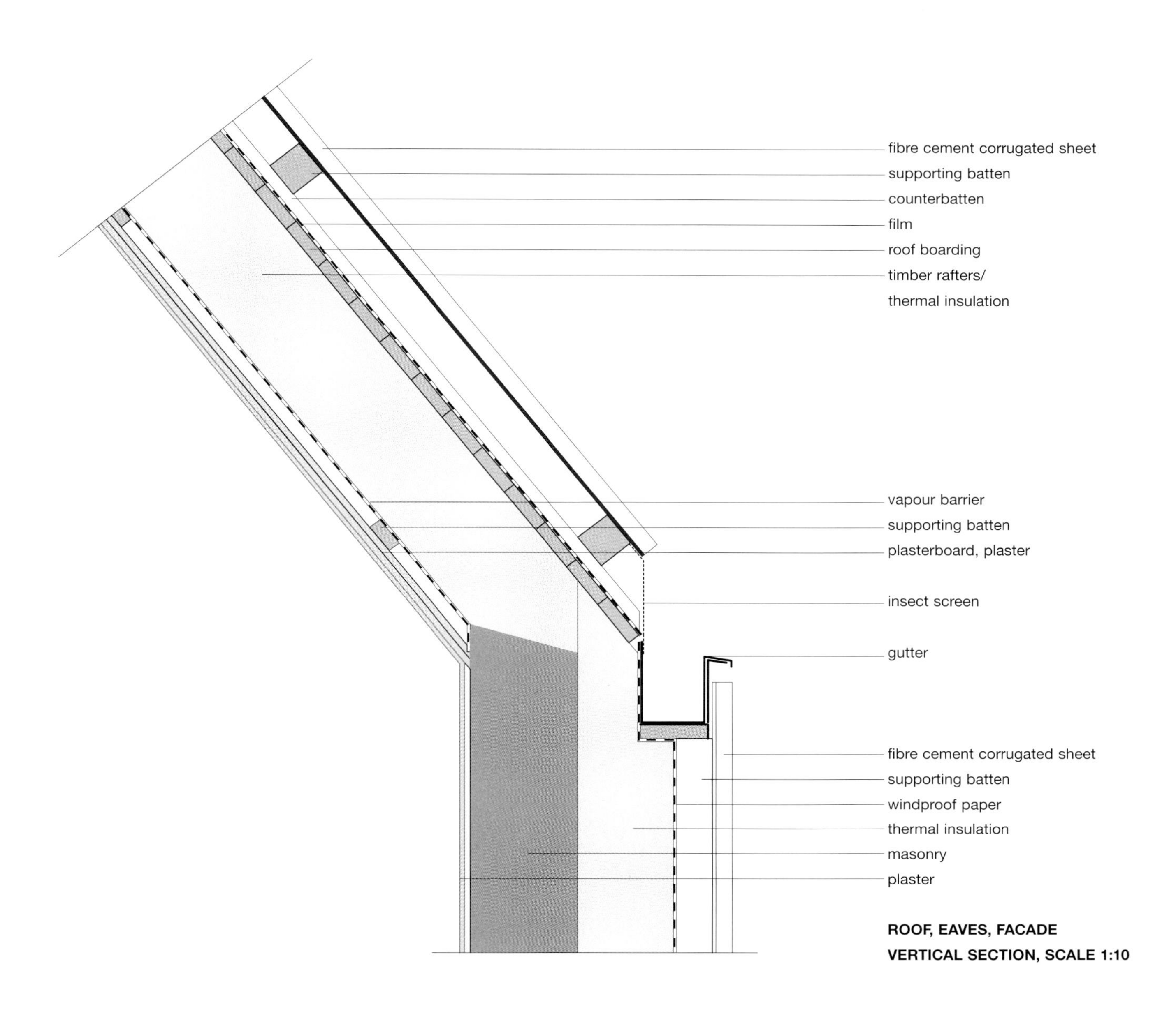

ROOF, EAVES, FACADE
VERTICAL SECTION, SCALE 1:10

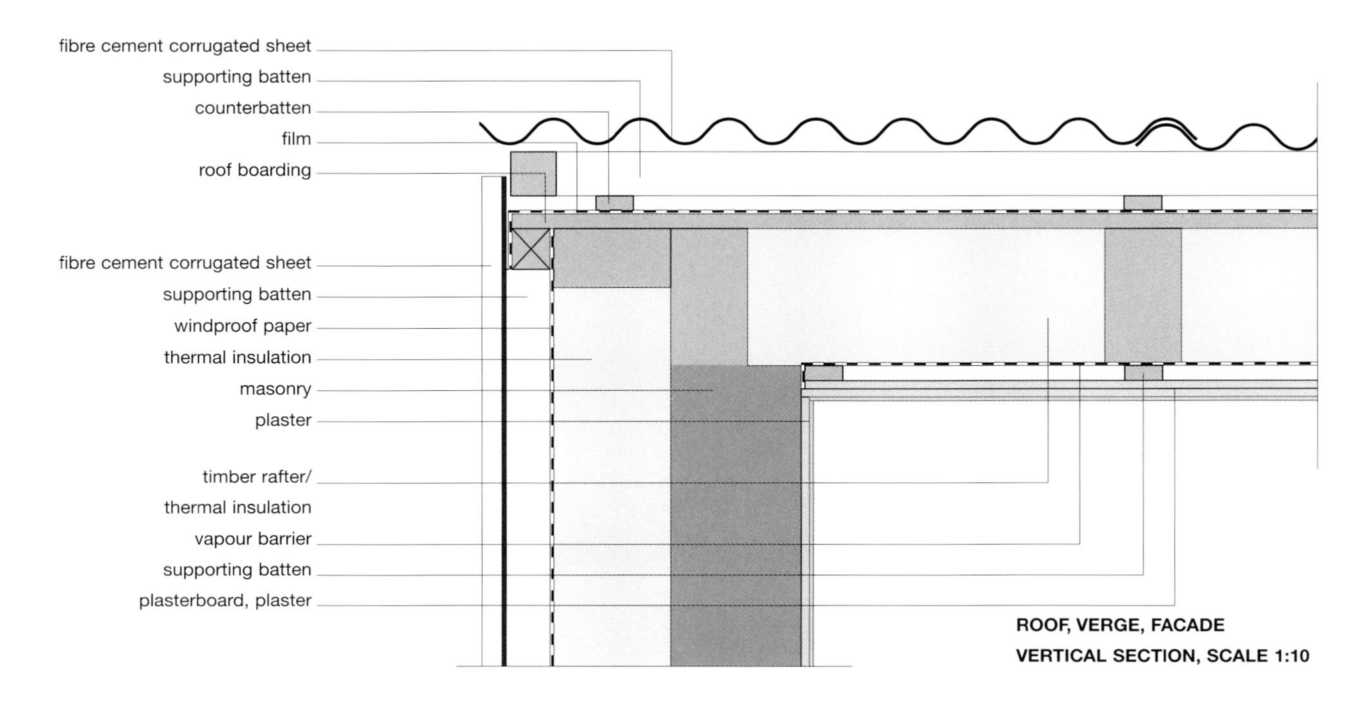

ROOF, VERGE, FACADE
VERTICAL SECTION, SCALE 1:10

Single-family house, Augsburg, Germany

Design: Bohn Architekten, Munich

Good architecture can only be created with strong clients who take up and implement the design idea of the architect consistently in every detail. The award panel for the German Architecture Prize judges outstanding works of contemporary architecture and recognises successful cooperation between clients and architects. This project was commended in 2003 and designed by the architect Julia Mang-Bohn as a completely normal house clad in black. The architect logically thought through the concept for the material and form. She specifically concentrated on a limited number of materials, making the most of each material so as to dispense with almost all additional elements. This also applied to the wooden structural frame of the house and the homogenous cladding of the roof and facades in fibre cement.
The traditional image of common lap shingles is immediately noticed on all four views of the house. The homogeneous choice of materials makes the roof the fifth facade. Black, 32 x 60cm fibre cement slates are fixed to laths and suspended from the substructure by slate hooks. The slate hooks are able not only to act as the structural fasteners but are also precisely positioned formal details, due to the fact that the exact fixing locations were on the elevation drawings. Unlike more solid roof tiles, the filigree shape of the 4mm thin shingles create a smooth, flat skin with the minimum of joints. Window and door openings are installed flush into this skin. Together with the dark cladding they become visually expressive wall and roof surfaces. The windows in the roof sit unusually low or high: on one side almost under the ridge, on the other directly at the eaves. The gutters are bent metal profiles, almost concealed from view and integrated into the roof, forming an elegant transition from the inclined to the vertical surface.

Precise Planning of roof and facade slates

Right: Smooth facade skin

Corner detail

Entrance set flush with facade

Dark bent aluminium profiles in the same colour as the fibre cement slates also create a fine transition to the verge and ridge. Instead of using a conventional preformed piece, which would have added a new graphical element, the ridge detail has sharp edges and was deliberately developed out of the original design idea. On the west side, which is exposed to the weather, the top fibre cement slates project slightly over the ridge. A long three-bend metal sheet is attached to the laths under the top slates to seal any holes on the opposite side. The side exposed to the weather is protected against wind-driven rain, whilst the substructure remains ventilated.

As with the roof, the facade windows are cut out of the fibre cement cladding as if with a knife. Large, fixed window bands combine with narrow, opening windows. On the gable sides the bands of fenestration extend to the verge. Cut at an angle, they are continued elegantly into the bottom corner and highlight the slender roof construction.

Indoor life contrasts powerfully with the dark fibre cement exterior. The high, open, light rooms are airy and generously dimensioned. The living room opens high, up to the underside of the ridge. This central space links the ground floor to the upper floor. The residents are passionate musicians and also use the living room as a stage with a large grand piano as the centre point. The stairs become the audience seating area for small family concerts.

These clever solutions extend beyond normal standards of work and required great craftsmanship. This precision lifts the house from being merely architecture to being a design object. The consistent use of materials in roof and facade brings the shape of the house into the foreground. So it attracts attention yet links with the character of the similarly shaped buildings in the neighbourhood on the edge of the city of Augsburg, between forest and cattle pastures. The dark clad, 45° double pitch roof is typical of the conventional single-family houses in the Swabian roof landscape. White gravel surfaces around the house and a light-coloured, exposed concrete pergola over the terrace form an elegant and contrasting frame for the unmistakable black shape of the house.

Photographs: Peter Frese, Wuppertal

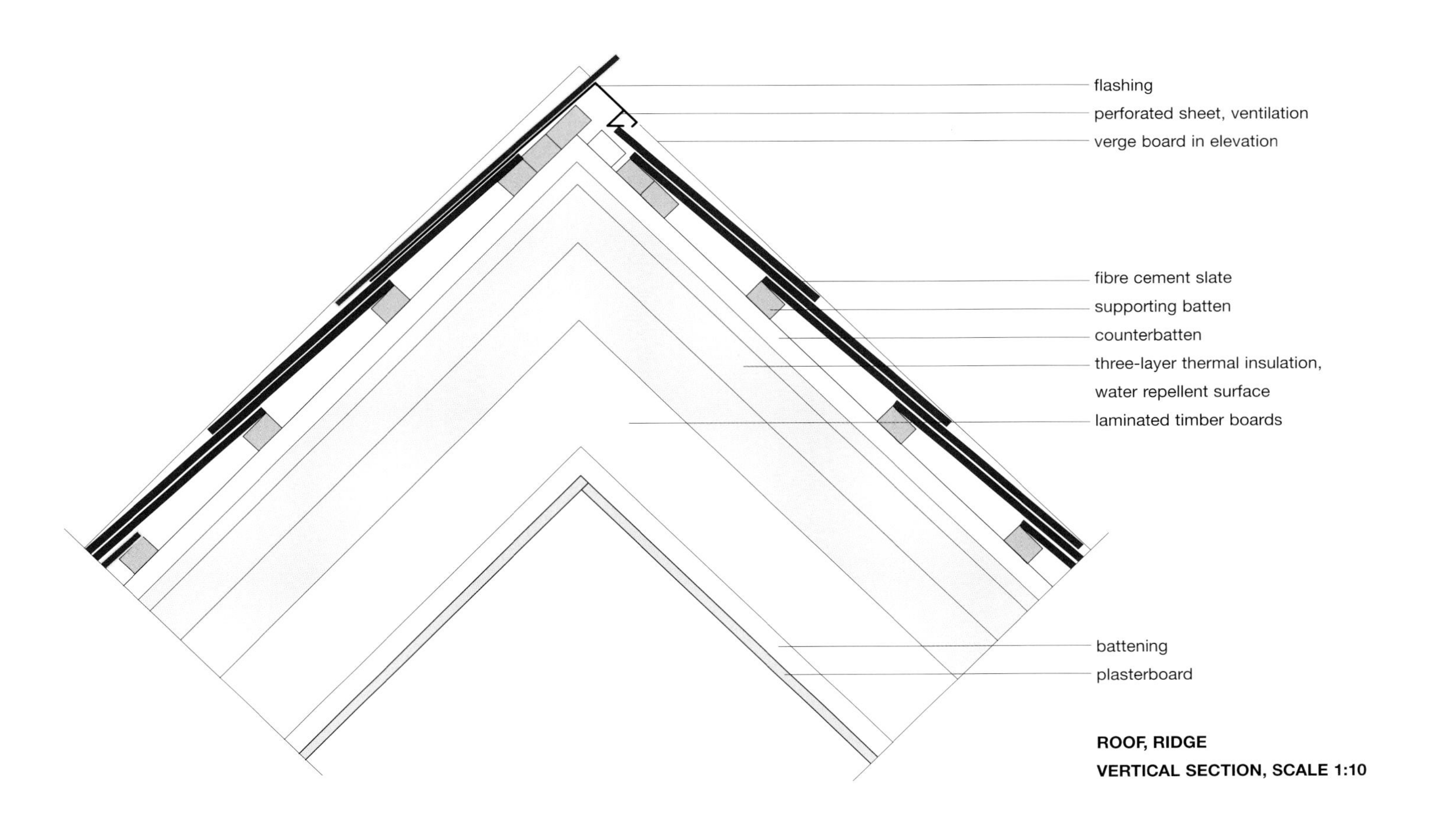

ROOF, RIDGE
VERTICAL SECTION, SCALE 1:10

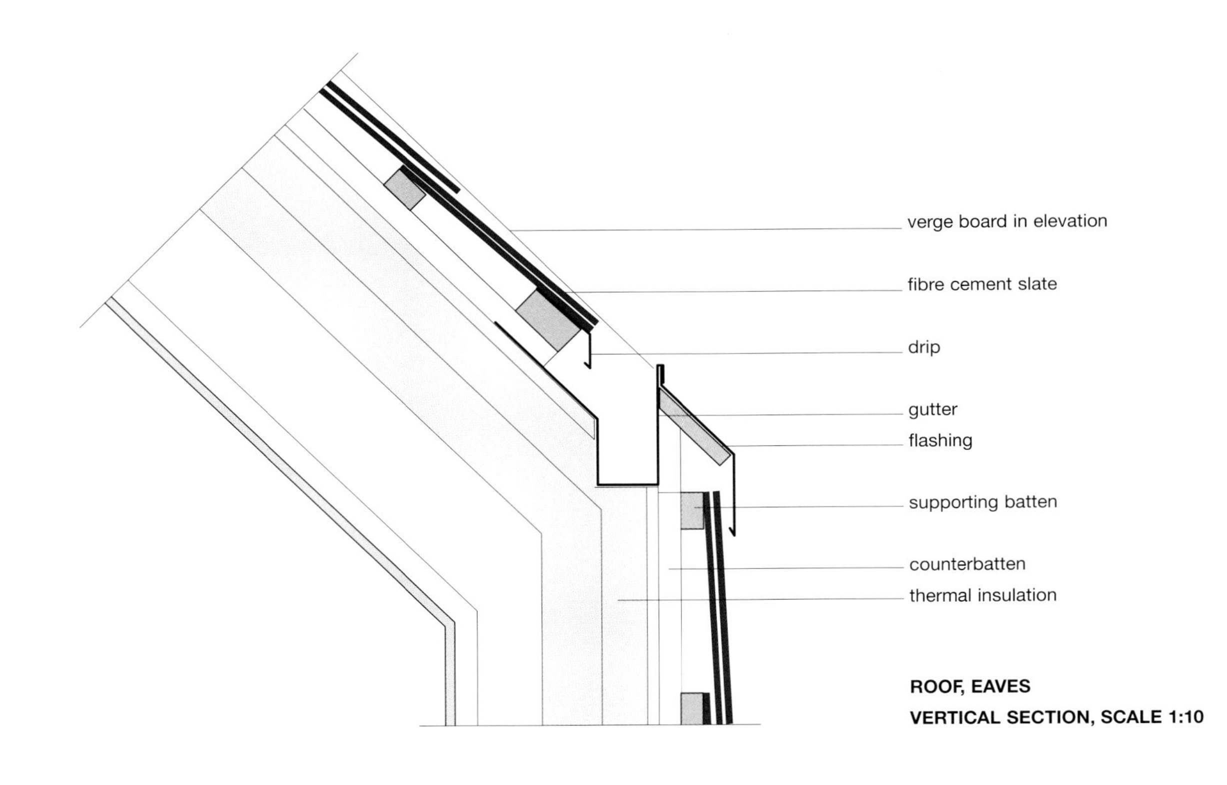

ROOF, EAVES
VERTICAL SECTION, SCALE 1:10

Commercial building in Norden, Germany

Design: Helmut Riemann, Lübeck

Large format fibre cement sheets for gable, longitudinal wall and roof cladding

This commercial office building consisting of a front wing with a street-facing gable and an elongated storehouse connected to its rear is typical of building developments in the small city of Norden on the North Sea coast of northwest Lower Saxony. A narrow alley separates the building from its neighbour. The front wing was stripped out and refurbished, but the store was so dilapidated that it had to be torn down. In its place Lübeck architect Helmut Riemann designed a new building, a task in which he elegantly perfects the balancing act between necessary adaptation and courageous innovation. Riemann has already converted, extended or refurbished several buildings in Norden and has made his mark on the contemporary image of the historic city. His designs show sensitivity for the substance of the buildings and their history. In the design process, he decides whether to adapt or to redesign, the signs of which can be seen in the finished building. His special attention to detail is remarkable in both design and execution.

A striking feature of the building is the bright red colour of the front wing, which dates back to the 1930s. The colour red was chosen because it could be used in various shades with other materials as a powerful and differentiating linking medium. The colour extends like a red thread from the facade on the old building through the lighting and the canopy materials and on to the facade of the new building.
The shape of the new building constructed on the site of the former store is similar to the old building. The designer's respect for the architecture of the city's historic warehouses with their characteristic brick facades is unmistakable. Shape and colour create the link, while the materials in the roof and facade provide an effective contrast, as the complete building consists of large format fibre cement sheets.
With a length of over 27 metres, the extension is much longer than the front wing. Its external proportions would tend to indicate only two storeys but a single elongated high room opens into the double pitch roof. Several timber frames placed at close intervals create an almost reverential atmosphere. The wall surfaces are decorated in light green, large parts of the roof and some of the facade facing the alley are glazed. The solid rear wall of the extension forms the counterpart to the open, rendered decorative facade of the old building. A narrow window opening in this wall promises a glimpse into the interior but is unreachable from the outside. Only when viewed more closely can the observer make out another opening: one of the facade sheets is actually a door.
The articulation of a building like this with a double pitch roof is normally evident at the transition points: from wall to roof or from wall to plinth, for example by a change of materials, with added elements such as gutters or balconies or by the recesses at windows and doors. It is quite different with this design: the uniform red lacquered fibre cement panels create a homogenous skin without projections and recesses, without added elements, an unbroken, shape-defining skin, extending apparently uninterrupted from wall to roof. The smooth flat surface is divided only by narrow joints between the rectangular sheets. The fibre cement sheets stand vertically over the full height of

Roof landscape with fibre cement roof

Material contrast: brick and fibre cement sheet

Generous spatial structure

Historic context for modern architecture

the storey and are precisely cut at the roof to follow the slope of the gable. Although the new building is scarcely higher than the brick buildings in its neighbourhood, its large format structure generates a homogenous, almost emblematic effect. The building, with its silky matt shimmering rust-red colour, provides a refreshing contrast to the strong brick-red of its neighbours, whilst giving a noble impression. The glazed openings close flush with the facade sheets and reinforce the impression of a smooth, flat, purely graphically articulated facade.

The simplicity of the homogenous appearance stands in stark contrast to the complex and skilfully executed details. The architect developed two systems for the different constructional requirements of wall and roof. The joints in the facades are highlighted in black and do not allow rainwater to enter. Behind the outer skin is the proven wall construction of a ventilated cladding facade: a layer of air separates the facade sheets from the building wall and insulation. In the roof and also at the ridge the joints are left open. The water is taken away by the underlying layer of bituminous roof waterproofing membrane. The panels on the roof have a purely formal function, whilst those on the wall provide a sealed protective layer. The potential problem of the transitions from one to the other is elegantly solved to achieve the impressive appearance of the building. At the transition between roof and facade the gutters are cleverly hidden by the sheets.

To reinforce the monolithic impression and the unbroken appearance of the skin, the roof and facade panels are attached using concealed fasteners. Aluminium agraffes are fastened to the recessed undercut anchors at the back of the fibre cement sheets and provide the structural connection of the sheets to the aluminium subconstruction.

The cladding, with its colour-coated fibre cement panels, creates a double pitch roofed building with an almost unique materiality. In this project the architect impressively extracts the essence from the building. The pure form, which is rediscovered only through the building's stringent unity of materials, is both restrained and fascinating.

Photographs: Lukas Roth, Cologne [4], Riemann [1]

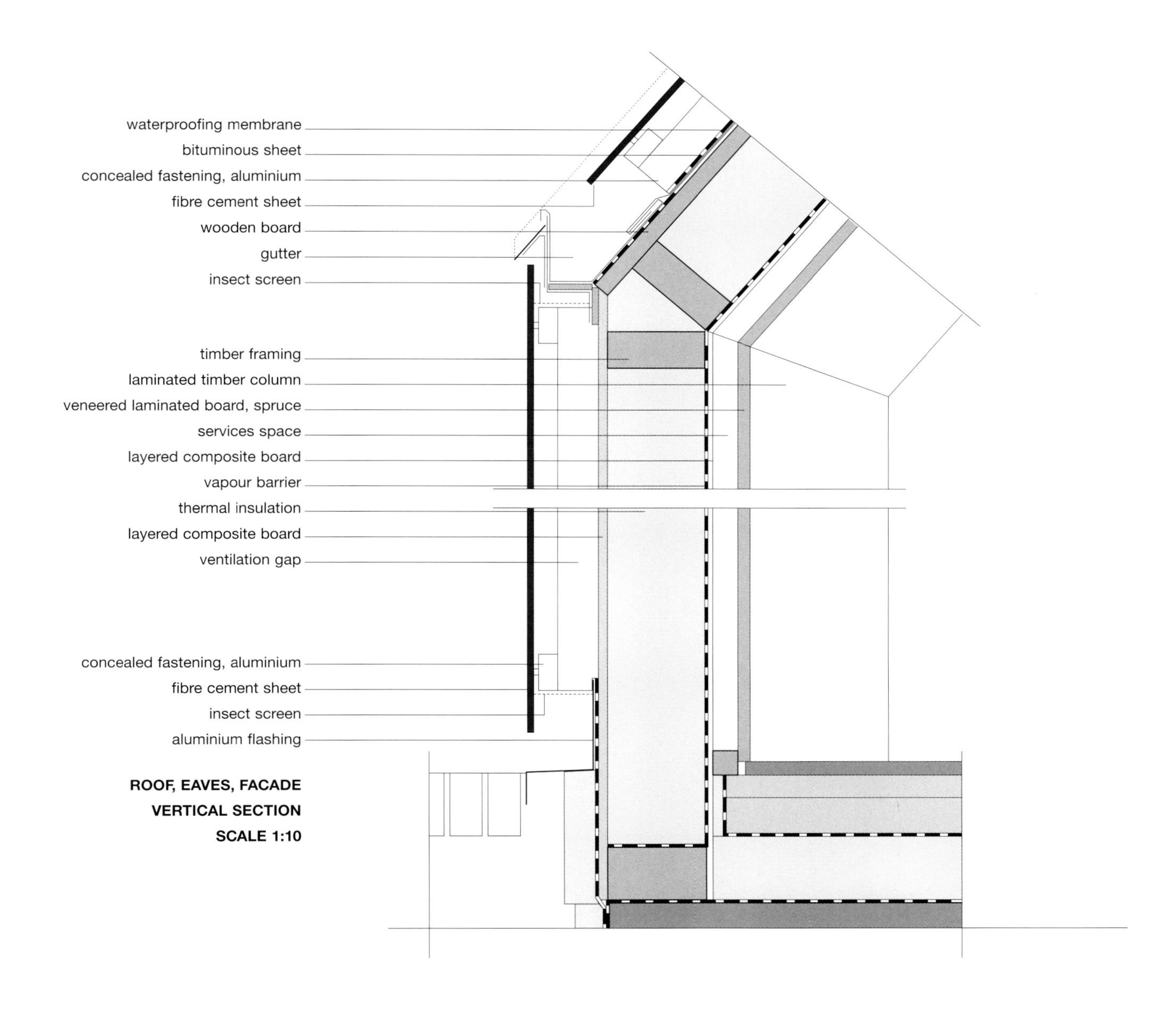

ROOF, EAVES, FACADE
VERTICAL SECTION
SCALE 1:10

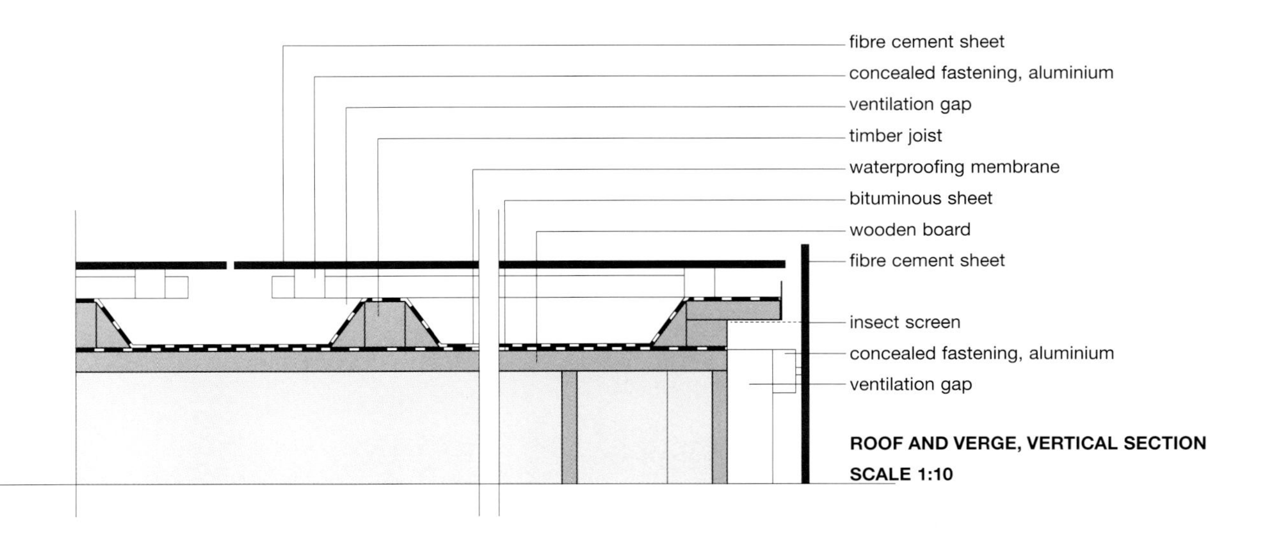

ROOF AND VERGE, VERTICAL SECTION
SCALE 1:10

Office building, Erpe-Mere, Belgium

Design: Christian Kieckens, Brussels

Atmospheric patio in the double gable

In the small town of Erpe-Mere in Belgium stands a somewhat different travel agency with the distinctive name of Caractère. Architect Christian Kieckens took the conventional buildings of the neighbourhood as the starting point for his design. He takes the form of a typical residential house with a double pitch roof, draws it out lengthways and doubles it in width. Then, as if the doubled building had been frozen in the process of drifting apart, the two gable wall peaks are surprisingly retained and intersect with one another, forming an M-shape. The double ridged building is only slightly taller than its neighbours, allowing the larger volume to remain in scale with the neighbourhood of houses, school and cloister.

The building is special not only because of its shape, but also because of its materiality and colour. The architect has clad the roof and facade with a single material: black, small format fibre cement slates characterise the homogenous skin of the building. Large, white-framed windows wander over the dark facades. They are not placed one above the other but are staggered by about two-thirds. On the ground floor of the elongated office building, three of the four corners have rectangular openings. One of these internally recessed openings accommodates the entrance. This entrance is painted white like the window frames and forms an effective contrast to the dark fibre cement shingle facade. The black skin is by no means monotonous but is rather more like a generous passe-partout for the definition of space, colour and materiality which lies behind the facade. No dormer windows, no roof windows, no penetrations interrupt the smooth, flat effect of the extensive, closed roof surfaces of the double gable roofed building. The architect came up with an elegant idea to allow the roof space to become an office for 25 employees: he fitted an internal courtyard between the roofs. This cut-out extends from ridge to ridge and provides light to the roof space from the centre of the building.

Rectangular, anthracite-coloured fibre cement slates clad the whole building. The roof material was extended into the facade to increase the homogenous effect. The scaly building skin has the effect of a made-to-measure suit. On the wall and roof, the small

Generous effect created by small format slates

32 x 45cm slates are fixed in rows and meet at the building corners and at the verge with a narrow joint. The principle used for the cladding can be seen here. The fibre cement slates are attached to the timber laths with simple slate hooks. In addition to their structural function, the hooks also give a rhythmic effect to the roof and facade. The straight-line arrangement of the sheets shows the particular precision exercised in the design and execution. Every projection, every slight curvature would be immediately noticeable on the close-fitting skin. Therefore the gutters are elegantly integrated into the pitch of the roof at the transition between roof and facade.
Simplicity and minimalism are not ends in themselves for Kieckens. This architect firmly believes that every addition to the architecture of a building should contribute to the content. His purism is not routine, as

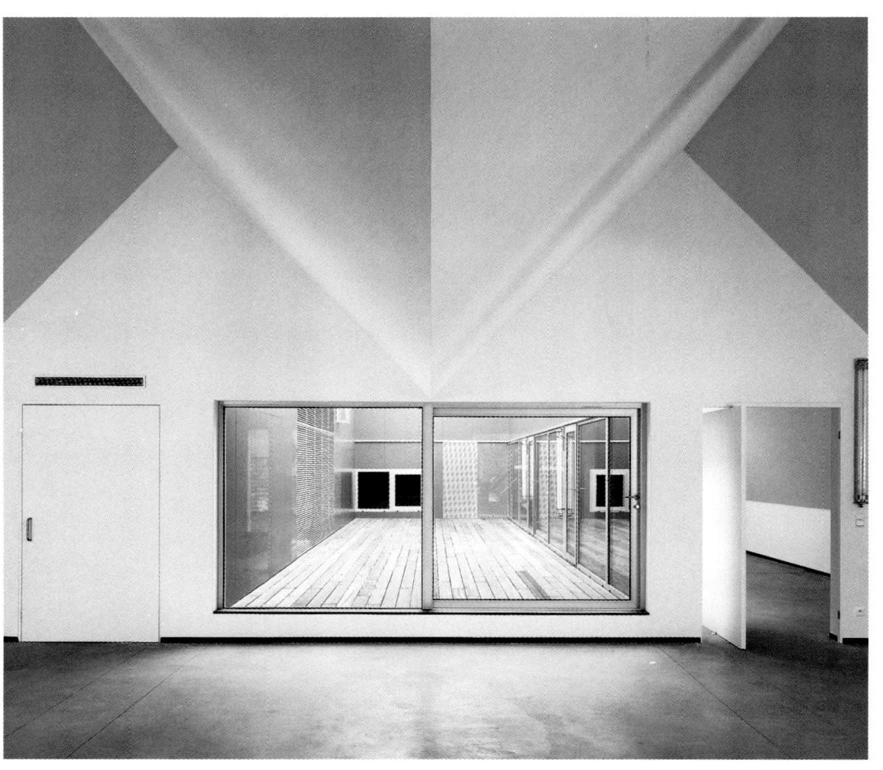

Inside, under the double gable

Precision in detail at the building edges

every decision is taken consciously: "I have no problem with another plane, but it should add some significance. Each additional input into the design has to make a contribution by conveying content, perhaps by asking a question or telling its own story." Therefore the architect first seeks to stay close to conventional trusted forms. His aim is not to exaggerate normality in an ironic manner but to gain new understanding with trusted tools. To achieve this he departs from the norm here and there and adds something special. The building satisfies the need for the familiar, whilst combining it with the curiosity about change.

Photographs: Rainer Lautwein, Bochum

Eaves with integrated gutter

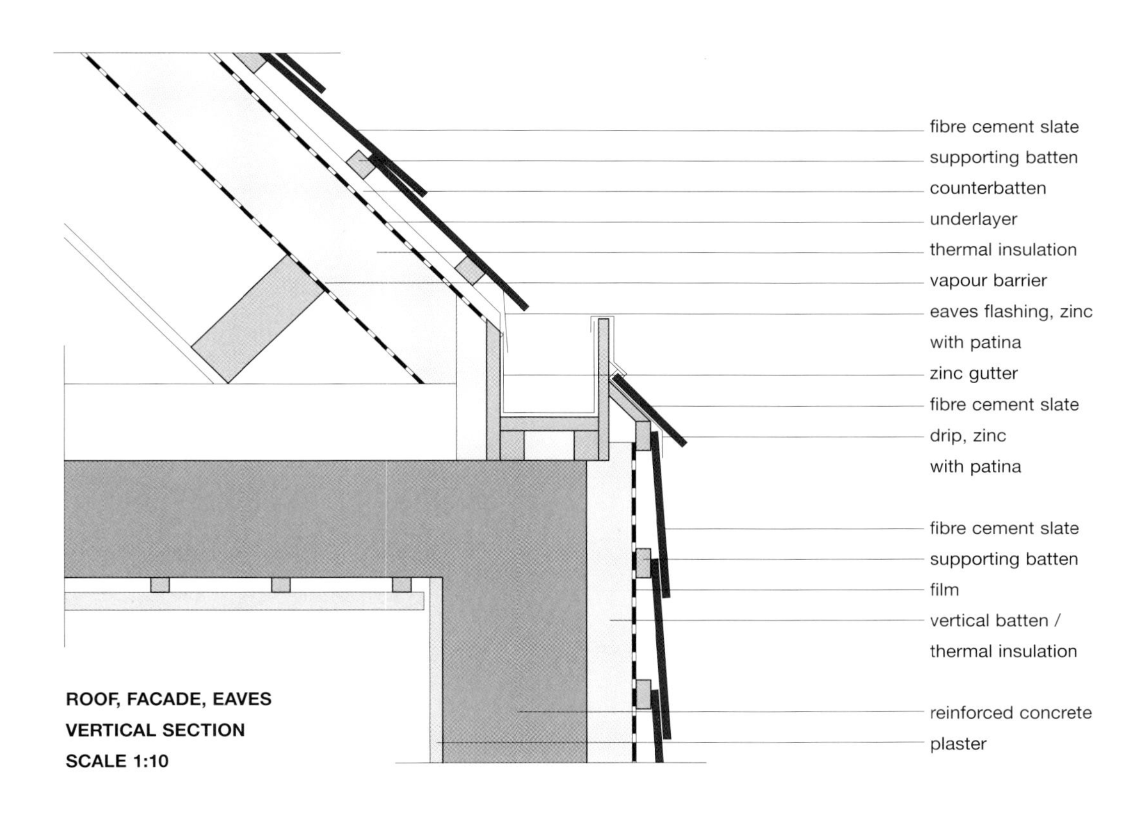

ROOF, FACADE, EAVES
VERTICAL SECTION
SCALE 1:10

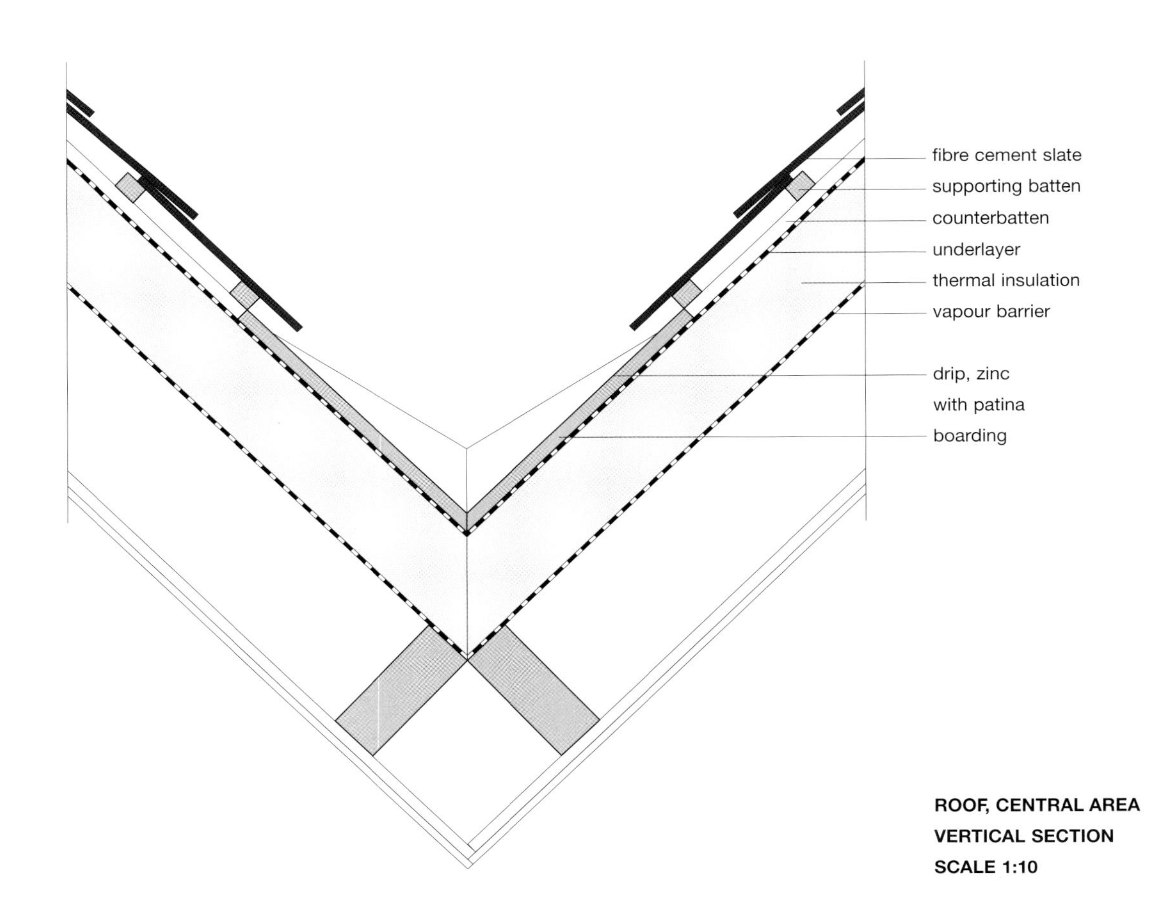

ROOF, CENTRAL AREA
VERTICAL SECTION
SCALE 1:10

Single-family house at Ely, England

Design: Mole Architects, Prickwillow

At the edge of a beet field on a street in the small English village of Prickwillow stands a black house. It is set in the countryside, the monotony of which is broken only by isolated farms and barns. The view wanders far over the fields. This part of east Cambridgeshire has a short history as an area of settlement. It was only 200 years ago that the fens were drained, resulting in one of the most fertile farming areas in England. Before this time, the nearby small town of Ely with its impressive cathedral was like an island of civilisation in a wide expanse of marshland.

House standing on base with entrance in gable wall

The Black House towers tall and proud above the flat landscape looking almost like a model to the observer. From ridge to raised plinth it balances loftily above the terrain. The entrance on the gable side of the three-storey house can be reached by a ramp or by steps. On entering the house it appears light and airy. Unpainted, naturally finished wood and white surfaces create a friendly atmosphere in the high rooms. The bathroom also enjoys uninterrupted views far across the fields. On the horizon the contours of Ely Cathedral can be made out on a distant hill.

The Black House took only eight months to build. In the initial stages, planning permission for this unusual proposal was refused time and time again: not because of the colour but mainly because of its height. The house has three storeys and stands on a plinth a good half metre above the ground. Many buildings in this boggy countryside have to stand on stilts to remain safe and dry during floods. The Black House has a plinth of glued laminated timber, which connects the structural frame of the building to the concrete piles of up to nine metres deep in its foundations. All that can be seen of these supporting works is the light limestone brick facing. These bricks are characteristic of the villages in this part of Cambridgeshire.

A further advantage of the building standing on a tall base is the improved air flow behind the ventilated cladding facade. Black fibre cement corrugated sheets clad the facade and the roof of the building. The zinc-coated profiles running in the horizontal joints between the sheets articulate the building. They do not mark the storey heights however, but are placed lower, directly above the window openings. The gable side is an exception. Large wooden-framed windows of different sizes stand out clearly from the black of the facade. They appear to be in rows because they all come up to the continuous metal profiles. Only in the top floor do the openings retreat into the background as black roof windows, so as not to interfere with the impression of protection given by the roof surfaces. The construction elements are prefabricated timber components. They are for the most part based on recycled materials, as is the insulation, which is made from newspaper. The relatively low weight of the tim-

Right: Roof and facade clad in black corrugated sheets

Profiles in the horizontal joints

ber frame construction cuts down the amount of concrete required in the foundations.

The barns in the surrounding countryside provided the inspiration for the Black House. Colour-coated fibre cement corrugated sheets are found on many agricultural buildings all around the region. As project architect and client, Meredith Bowles made a conscious decision to design the building as a home for his young family. Meredith Bowles believes that making reference to the local architecture leads to the building fitting better into its surroundings: it is "a region with its own strong identity". In addition the architect sought to reinterpret familiar motifs that would be free of the usual country house clichés. Fibre cement corrugated sheets had already been used in a similar context in Denmark and Holland to reflect the rural architecture there. In Great Britain a new approach was developed. "It demonstrated that a relatively inexpensive product can, with a little effort, be turned into something fairly extraordinary. Good-looking details can be fashioned from standard elements and accessories with some ingenuity. The effects created as the light strikes the profiled building skin are magical," enthuses the architect. Depending on the viewing angle and the position of the sun, the facade looks striped with a certain depth or as a matt, shimmering, apparently smooth surface. The light wooden window frames and colourful doors take away any gloominess from the black skin. The image of the building towering straight up is reinforced by the vertical orientation of the corrugations on the fibre cement facade. Bowles explains about the material: "It is more durable than steel as corrosion is not a problem. And if necessary it can be cleaned and another coat of paint applied. The effects created by the profile on the vertical surfaces look fantastic."

The wall construction is very simple. Standard fibre-cement corrugated sheets give the building an impressive profile. Roof and walls are similar in construction. A metal substructure connects the corrugated sheets to the supporting wooden frame. There is no overlap where the top and bottom of adjacent corrugated sheets meet. A continuous horizontal profile is used instead. The cover plate beneath the gutters at the eaves corresponds with the horizontal joints between the corrugated sheets. A striking corrugated ridge cap completes the double pitch roof. Just as distinctive, the corners of the building are rounded: the downpipes are appropriately integrated into the facade.

The same characteristics that made fibre cement first and foremost a functional material in industrial and agricultural buildings are now valued in other contexts. The material is easy to maintain, durable, robust, and permeable to water vapour. In combination with other materials and highly contrasting arrangements of coloured elements, the status of what used to be a purely utilitarian material is enhanced. The black building with its innovative facade is attracting a great deal of attention – not just in the neighbourhood, but also in national architectural circles and beyond. The Black House proved its special character by winning a national architectural award in a competition organised by the Royal Institute of British Architects (RIBA). The house successfully combines innovation and tradition through sensitive use of the material in the context of its well thought-out design concept. The rather bland tone of the countryside is enriched by the contrast provided by this stark yet lively building.

Photographs: John Donat, Hong Kong

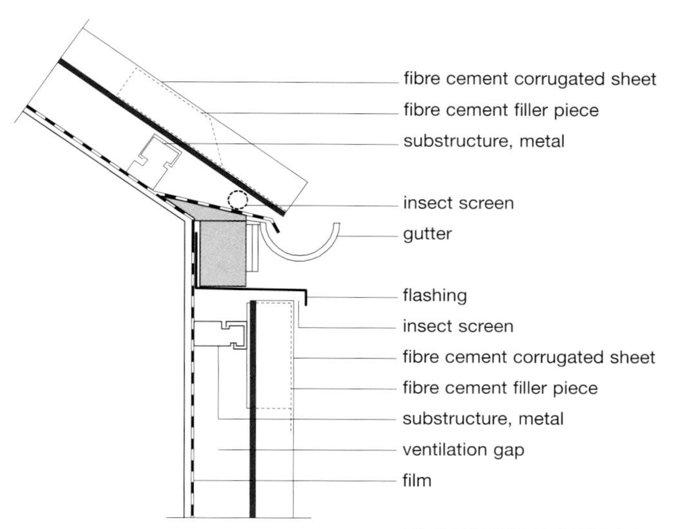

ROOF, VERTICAL SECTION
SCALE 1:10

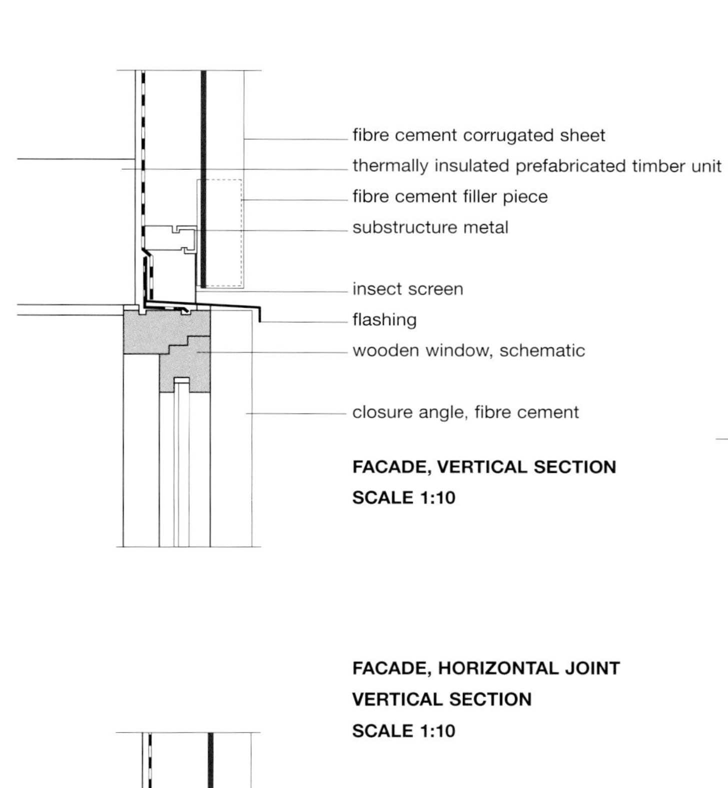

FACADE, VERTICAL SECTION
SCALE 1:10

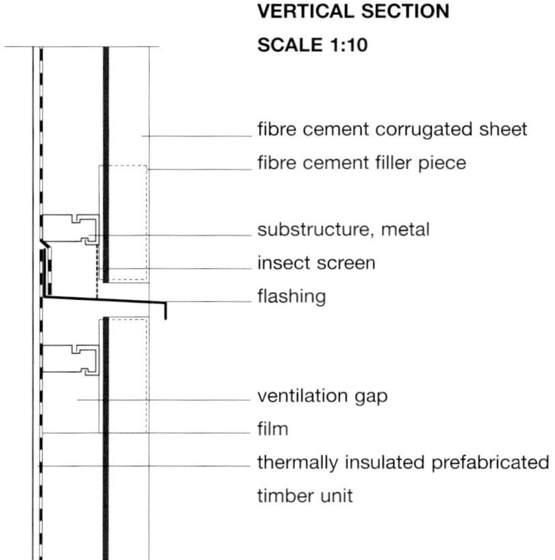

FACADE, HORIZONTAL JOINT
VERTICAL SECTION
SCALE 1:10

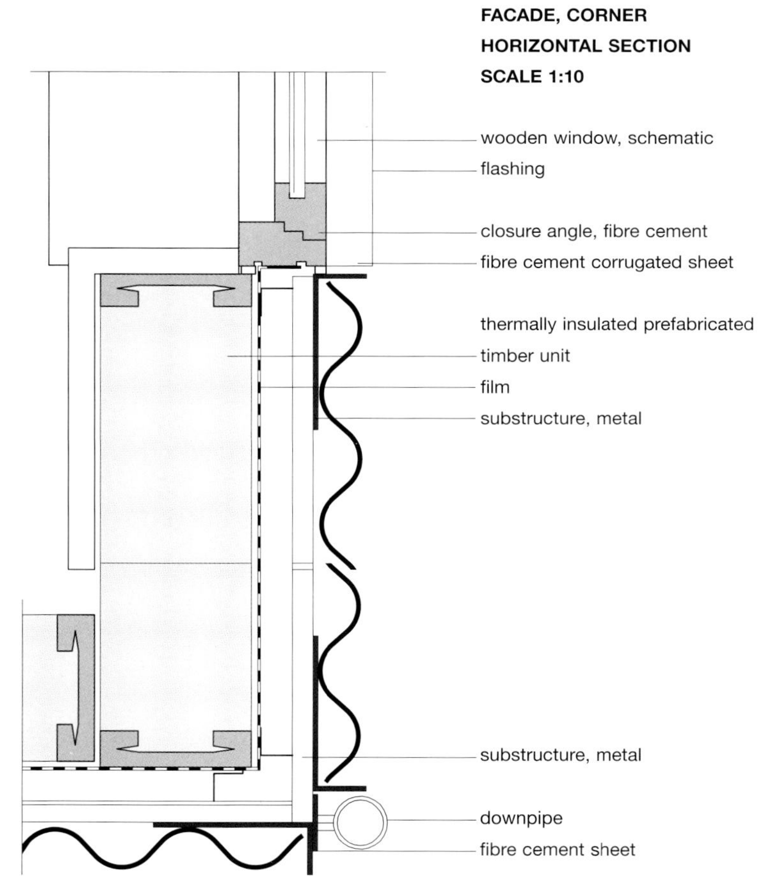

FACADE, CORNER
HORIZONTAL SECTION
SCALE 1:10

Residential building, Frankfurt am Main, Germany

Design: MMZ Architekten, Frankfurt

Large format fibre cement strips in the roof

In a typical area of new development in Frankfurt am Main stands a huge, red version of a single-family house. Whilst complying with the building authority's requirements, the adapted form of this detached house, with its double pitch roof, red cladding and light render, is quite different from its neighbours. It dispenses with any overhangs and projections, including eaves projections, as the angular monolith is reduced down to its archetypical basic form. On closer inspection of the small details and consideration of the building as a whole, it becomes clear just what qualities this apparent simplicity hides.

Architects Claus Marzluf, Ray Maschita and Thomas Zürcher originally planned for a four-family dwelling, but today the house is occupied by only one family. The reason is that the client decided to move in himself with his family and office. Very few changes had to be made to the original design to accommodate this new use. No modifications to the openings in the facade were necessary as it was never intended for the apartments to be discernable from the outside. From the beginning, the building always had the clear shape of a typical single-family house – only on a much larger scale. At almost 14m high it soars effortlessly over its neighbours.

The appearance of the house is greatly influenced by the consistent unity of its surfaces. Wall and roof flow into one another seamlessly. The building uses two completely different materials and systems of construction: the walls are rendered, whilst on the other hand the roof is clad with large format through-coloured fibre cement sheets and forms a uniform plane. The almost identically-coloured surfaces of the roof and walls create the perfect transition and combine themselves into a single skin. All elements of the building are co-ordinated with one another and woven together masterfully. The fibre cement clad chimneys rise up from the strips of similar material on the roof and their tops are cut to match the 45° angle of the opposite roof slope. The copper gutters are integrated and concealed in the roof surfaces, so that the eaves look as sharp edged as the verges and appear as clear lines.

The special effect of the house comes from the design of the roof, where the architects rejected the small format roof coverings, such as the clay or concrete tiles commonly found in the neighbourhood, and instead chose large format fibre cement sheets. The 8mm thick, rectangular fibre cement sheets are placed in sizes of 30 x 150cm and 30 x 180cm over the surface. They are glued in a double layer on to an aluminium substructure. The principle is the same as that used for the ventilated cladding facade, but it is sloped at 45° from the horizontal. The fibre cement sheets have a purely formal function. The real function of a roof is performed by the underlying waterproofing layer. This roof liner is sealed using a membrane in the same way as a flat roof but at a slope of 45° from the vertical. The clever combination of flat roof and ventilated facade, with the clear and consistent separation of the functions in two layers, is the constructional secret behind the aesthetic straight lines.

Homogeneity from vertical bands of fenestration

Sharp-edged gable

Spacious interior

Eight vertical window strips on the long sides of the building extend from the ground up the facade and eaves and into the roof, thus reinforcing the homogeneity of roof and walls. They open up the body of the building with their regular pattern. Windows alternate with copper sheet facings. In the course of time the copper, which is also used in the window spandrels, will take on a patina and add another colour to the building.
The house and the interiors strike a balance between being a design object and large-scale architecture. Careful detailing is evident right down to the finishing touches and effectively underlines the basic concept of the house in its pure form.

Photographs: Jörg Hempel, Aachen

Between design object and large-scale architecture

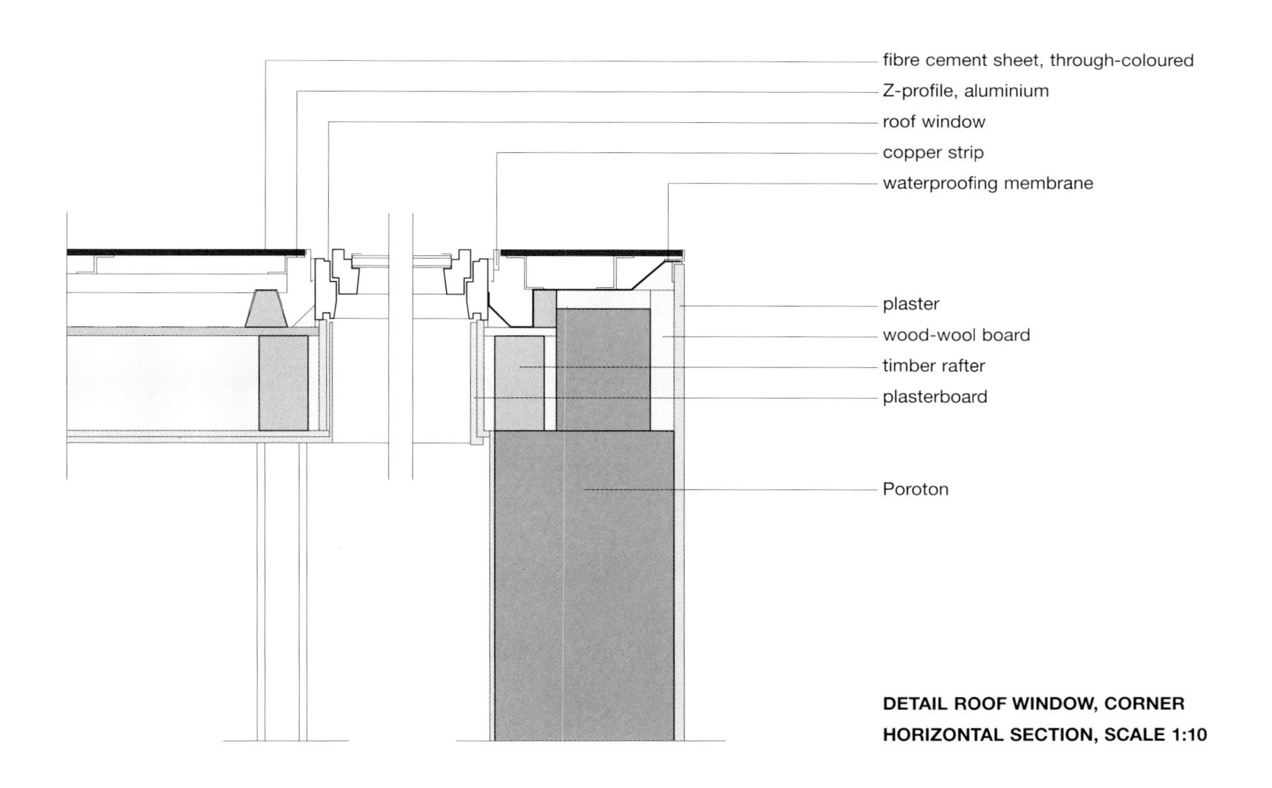

DETAIL ROOF WINDOW, CORNER
HORIZONTAL SECTION, SCALE 1:10

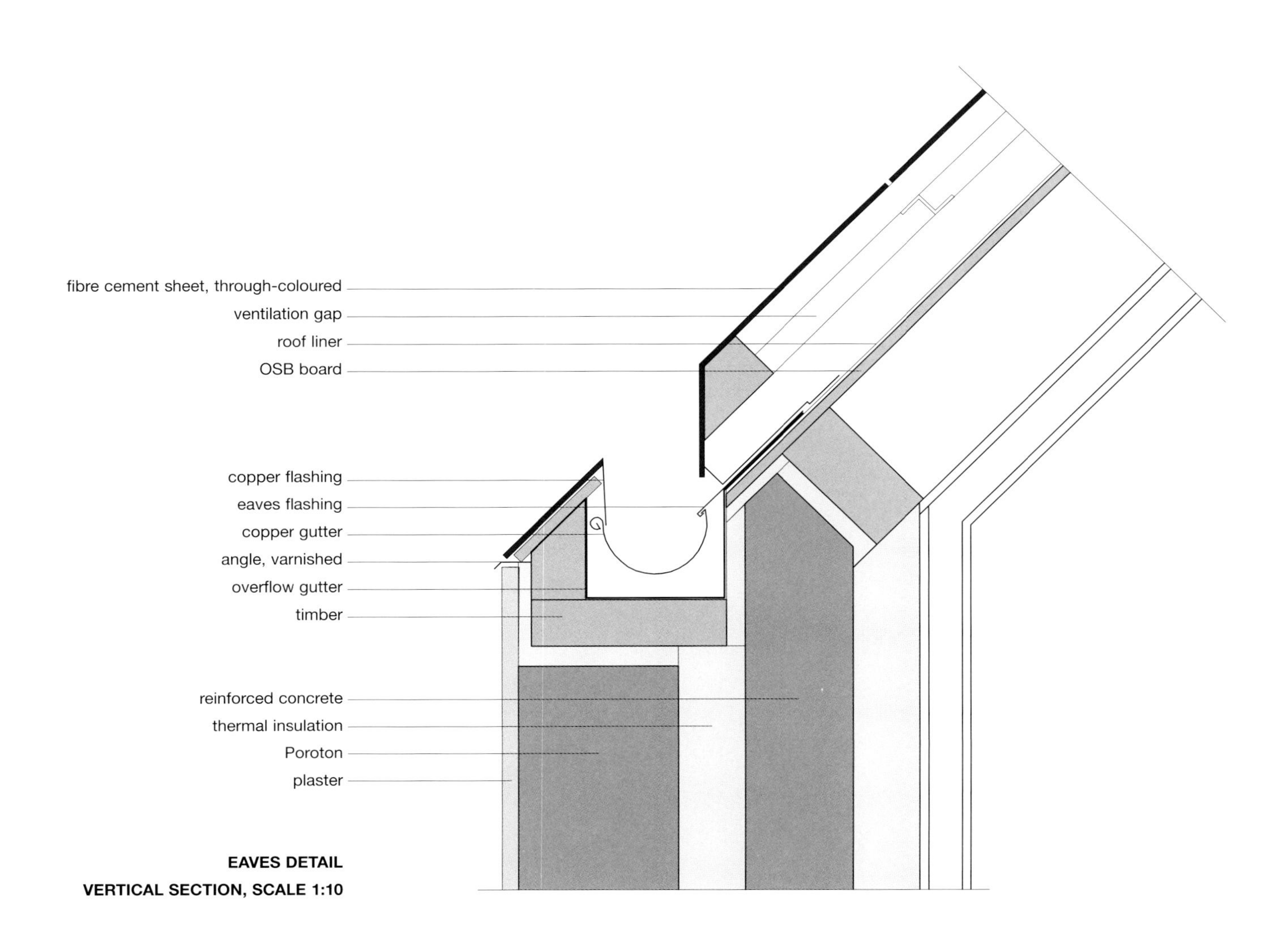

EAVES DETAIL
VERTICAL SECTION, SCALE 1:10

Single-family house, Randalstown, Ireland

Design: Alan Jones Architects, Belfast

Woven pattern of anthracite fibre cement slates

The house where the four-strong Jones family live in Northern Irish Randalstown stands on an unusual piece of ground. The access road leads between the cemetery and the Orange Hall, a Protestant community meeting place. Its neighbour directly opposite is a listed 18th century Presbyterian church. The church is oval in floor plan, constructed from brown basalt random rubble masonry with a slate roof and hence is somewhat unusual. It seemed a natural reflex to the architect Alan Jones for the design of his own house to be calm and reserved. No coloured or bright elements were to distract from the view of the church. And yet it should not look like a private dwelling, as the context demanded a building with a higher standard of urban design. "This site needed another seemingly public building. That gave us the opportunity to build big, be ambiguous and omit the details that a private house normally demands," says the architect.

Black box behind the graveyard

The choice of material for the roof and facade helps to ensure that the house can assert itself despite the proximity of the church. The whole building is homogenous and clad with almost completely black fibre cement slates. The blue-grey through-coloured shingles on the facade form a textile pattern. The architect selected 60 x 60cm slates, a particularly large format, to create the right appearance for the open lap facade cladding. The laterally overlapping slates are held at the bottom edge by two flat hooks. The large format panels emphasise the urban look of the building following its solidly built, civic neighbours.
Depending on the viewing direction along the surface, the line of the edges of the sheets forms an up or down zigzag pattern. The black skin of the building is in accord with the dark slate roof of the church and some of the neighbouring buildings. The architect achieves a modern contemporary interpretation of a deeply traditional motif on this site. The way in which roofing material has been put on a facade is unusual. At the windows too, the facade can still be perceived as a suspended, non-load-bearing element. Each slate is accurately trimmed at the corners of the reveals, as the window openings are not the same size as the slates. This reinforces the impression of a homogenous skin. The dark, closed facade is free of interference from attachments such as gutters and downpipes. The proximity of several deciduous trees means that rainwater from the roof and walls can be allowed to flow directly into a gravel bed surrounding the building. Like the facade, the roof is predominantly a closed surface: no chimneys, no dormer windows, nothing to disturb its flat, smooth appearance. Only a few roof windows are inconspicuously inserted into the inclined black plane.
Colour and texture iridesce on the character-rich pattern of the wall cladding. When the high sun shines directly onto the skin, the black, scaly surface appears almost smooth with a silvery grey, metallic gloss. Evening daylight casts long shadows between the individual shingles, which then look like thick stone slabs. The matt dark grey of the fibre cement slates reveals its delicate texture only at close quarters. Visitors to the church perceive the facade of the house

Light and shadow relief on the gable wall

Homogeneous building envelope

at some distance. From this far away the woven surface texture sometimes subsides into a shadowy blackness. Cloaked in its dark scales, the house looks mystical and full of secrets.

Photographs: Alan Jones Architects, Belfast

Scaly surface with a metallic gloss

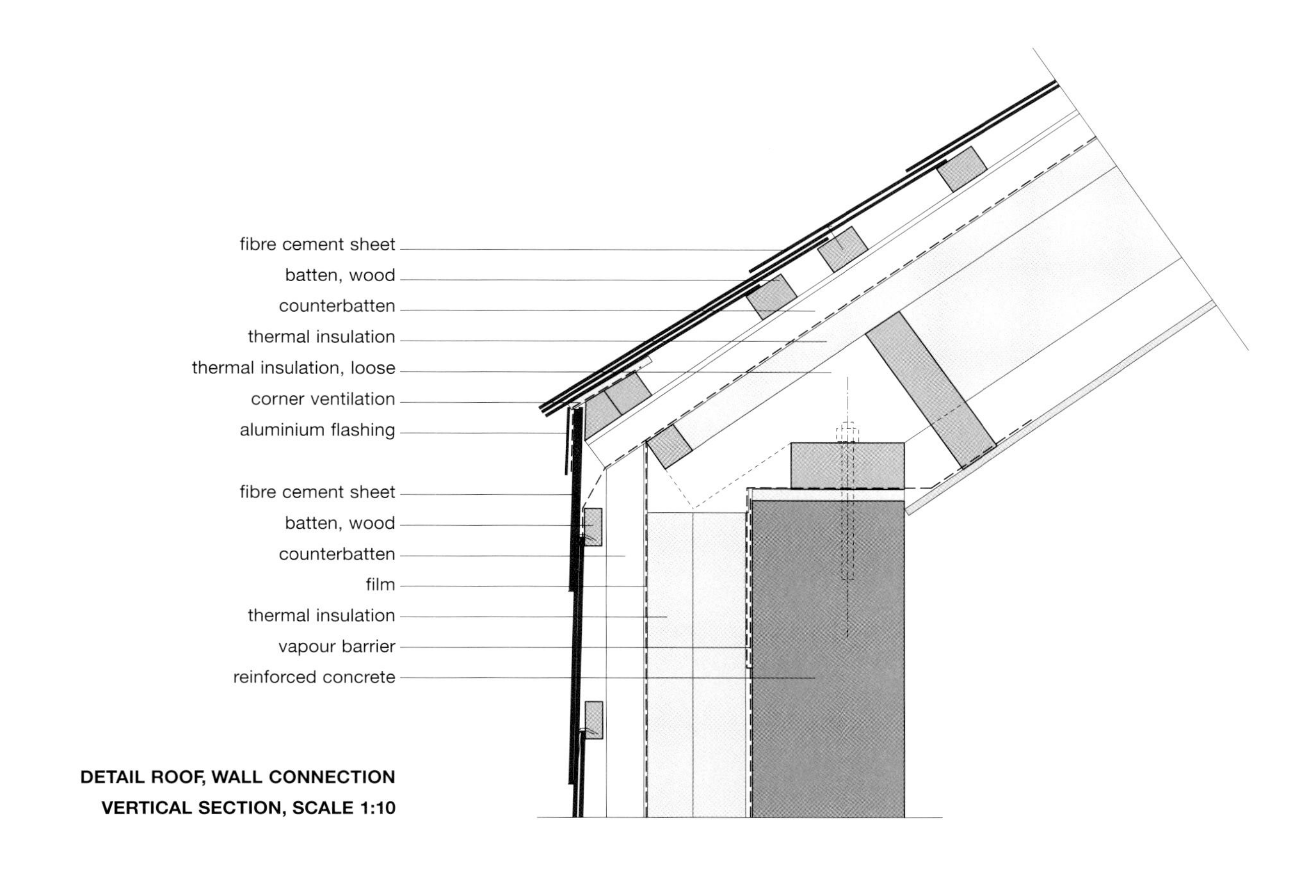

DETAIL ROOF, WALL CONNECTION
VERTICAL SECTION, SCALE 1:10

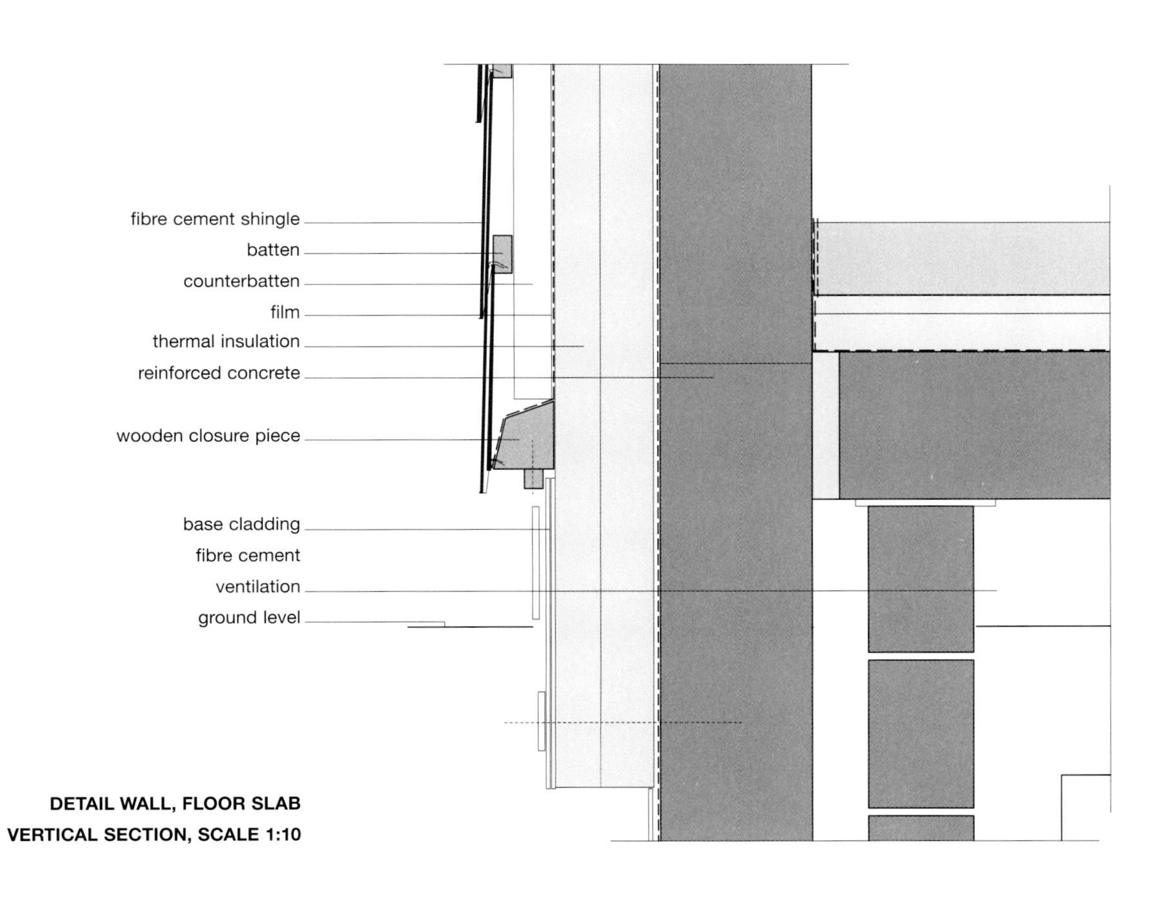

DETAIL WALL, FLOOR SLAB
VERTICAL SECTION, SCALE 1:10

Interiors with fibre cement

The substantiality of a material is a major and form-defining design attribute in contemporary architecture. Colour, structure, texture and haptic are the principle levels of perception that fashion a new consciousness of space and ambience. A material's identity, authenticity and the sensuality of its surface play a special role in interior space. Architects have always sought to exploit materials in unfamiliar ways, for example, using materials intended for exteriors on the insides of their buildings and visa versa. An obvious motif is to bring the facade material into the interior space and shift the boundary between inside and outside. Within buildings as well, even those without fibre cement in roof or facade, this material can be found in their interior architecture, often in combination with glass, stainless steel, wood or stone. Indoors the material reveals its own special sensuousness, unfolding a haptic dimension. It is made clear again how little the material is defined for any particular purpose. The selected examples show a range of different interpretations of the same material for walls, ceilings and furniture in a showroom, bar and museum. A piece of fibre cement furniture as a monolithic block becomes a room-defining element, whilst as a thin veneer, the cement material offers a reserved passe-partout.

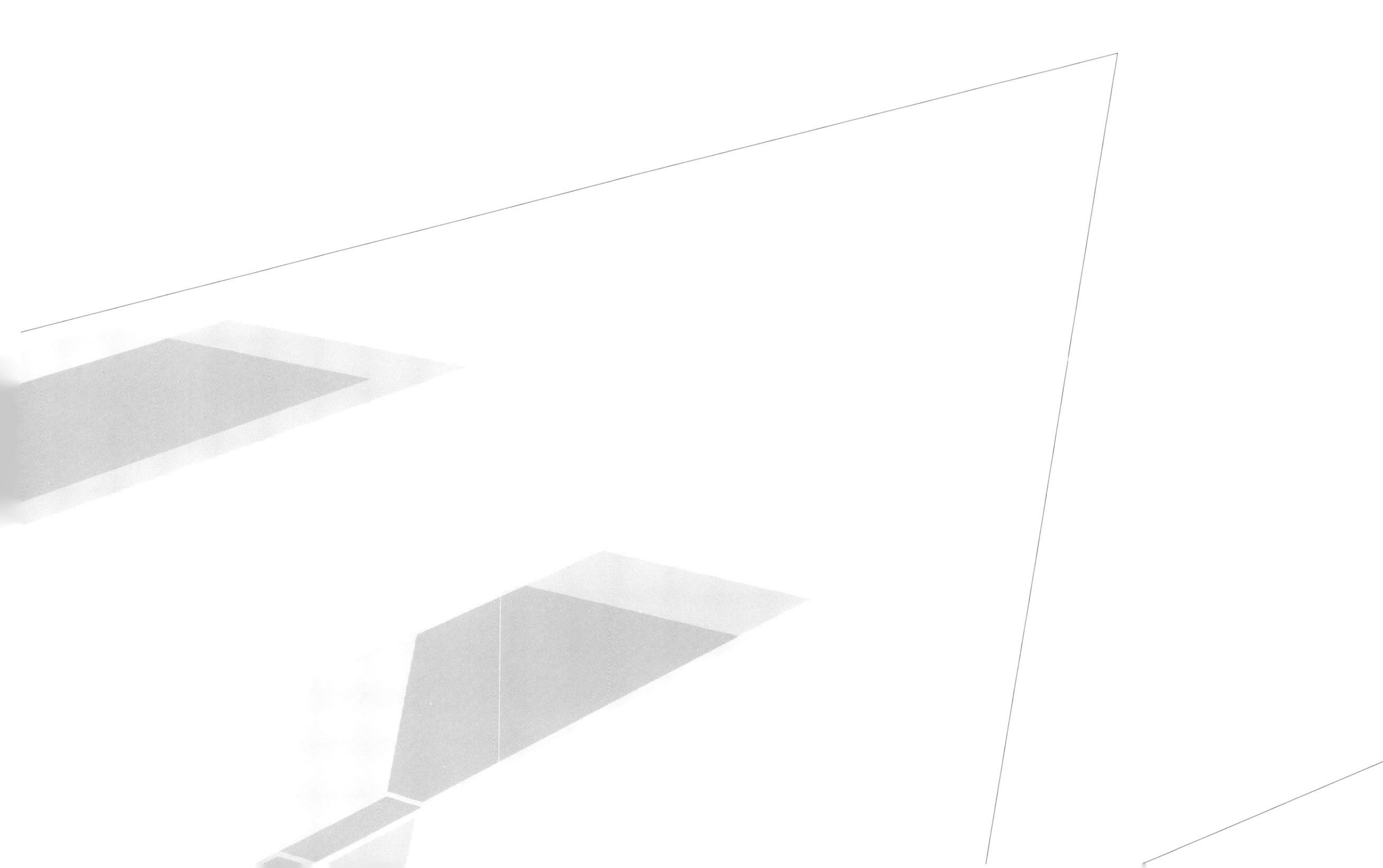

Showroom and training centre, Heidelberg, Germany

Design: Astrid Bornheim, Berlin

Spacial dynamic with fibre cement furniture

Architect Astrid Bornheim thinks of the Showroom and Training Centre at Eternit AG in Heidelberg as a "laboratory for ideas, a workshop for inventions and architecture for activity spaces". At the heart of the design is the idea of using fibre cement as a space-defining element. At the reception it can already be seen what is depicted in the subsequent series of rooms: framed views into and out of the rooms reveal their interior form. Fibre cement is the material used as the frame for communication and action, and is the stage, main screen, passe-partout and exhibit, all at once.

The Showroom and Training Centre is housed in a culturally and historically important building. It was built in 1964 by the architect Ernst Neufert and is now the company's administrative headquarters. The clarity of the design of the newly configured rooms retains Neufert's architectonic structure. The construction and proportions of the rooms, in which poorly executed alterations had been carried out over the years, were restored. The finely detailed concrete ribbed slabs of the structural frame give rhythm to the various new planes and elements. A room-high glass facade plays

Right: Fibre cement as exhibit and as room divider

Lending rhythm to the room

around and escapes the set structure. The internal relief wall provides a second, contrasting motif. The room in between develop each a rhythm of their own by the variously sized fibre cement furniture.
The architect developed a compact, multifunctional room arrangement, which defines the spaces for commercial meetings, exhibition areas and for formal and informal discussions. The room also has free-standing blocks of different sizes made from fibre cement. They fulfil the simultaneous functions of exhibits and room dividers, reception desks, display stands and espresso bar. This fibre cement furniture fashions different functional zones and yet always allows the visitor a glimpse of the whole. The furniture divides and connects. A series of inserted rectangular showcases opens up views from the entrance to the bar.

Material combination: fibre cement and glass

The client followed the recommendations of the architect not to present products on a conventional exhibition stand but to invite the visitor to discover them. The exhibits – corrugated sheets, facade sheets, concrete roof tiles – can be found in fibre cement containers, which are integrated in a natural way into the architecture of the room. The design permits the visitor to explore the haptic qualities of the material and creates in the truest sense of the word points of contact and multilayered opportunities for product identification.
The only visible material openly displayed is the basic material from which all the other products are made: 8mm thick natural grey fibre cement sheets. The architect uses it with an extraordinary complexity and universality: as wall coverings, door leaves and a table-

Fibre cement for tables, doors and relief wall

Inclined glass plate forming wash basin in the bar counter

top in the seminar room. The sheets, up to 3100mm high and 1250mm wide, are glued to a substrate of MDF boards. They were cut to the shapes detailed in the CAD drawings using water-jet cutting, which allowed the sheets to be cut to any shape or have any shape cut out of them. The fine linear pattern of the narrow 3mm joints called for extraordinary precision from the craftsmen when fitting the sheets.

The design motif of discovery is also emphasised by the combination of the latest materials and techniques. Display cases and balustrade elements are fitted with special Japanese glass that obscures fleeting views and only reveals an exhibit to the observer looking directly at right angles to it.

The light, with its buoyancy, follows the rhythm set by Neufert. Here too, the architect selects a new product: a slim glass rod combines the roles of lighting tube and halogen spotlight. The colour accord of the room is taken from the company logo: the red and green colours of the company logo appear in glasses, chairs and tableware which are displayed against the background of the white, jointless, epoxy resin floor and the grey passe-partout of the fibre cement furniture. At the bar the counter is panelled with fibre cement and topped with green glasses. The wash basins are formed from inclined green glass plates in a wide stainless steel channel. The fibre cement material is protected from moisture this way and forms the robust, characterising core of the wet areas.

What Ernst Neufert demonstrated in 1964 still applies to the new entrance today: fibre cement can be used in an experimental way to a remarkable extent and formulated in new contexts. Whilst showing full respect to the Neufert heritage, the architect demonstrates a new poetic dimension of the grey construction material and provides the company identity with its visible built expression.

Photographs: David Franck, Stuttgart

Multifunctional room arrangement: reception, display stand and espresso bar

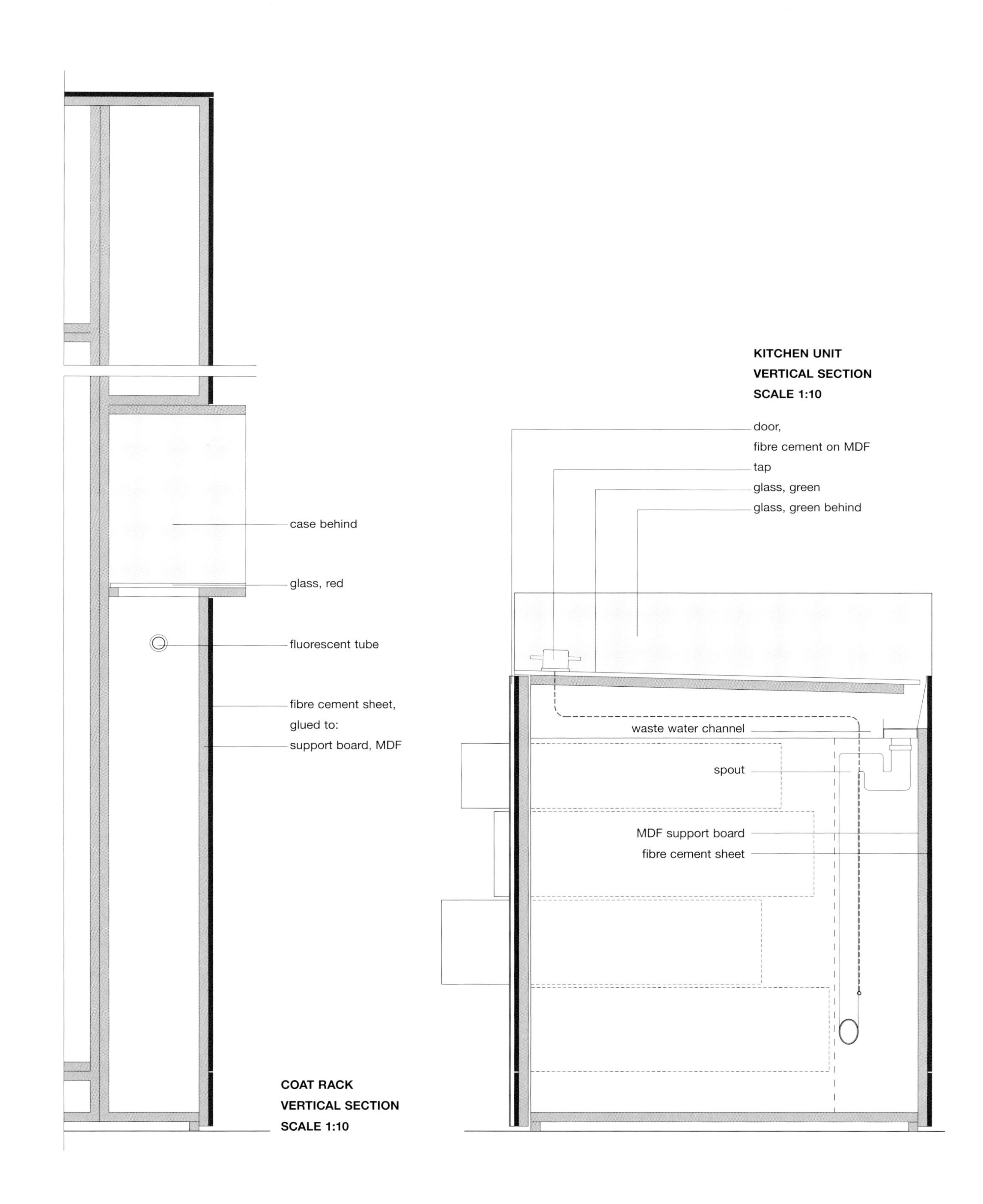
case behind
glass, red
fluorescent tube
fibre cement sheet,
glued to:
support board, MDF
COAT RACK
VERTICAL SECTION
SCALE 1:10
KITCHEN UNIT
VERTICAL SECTION
SCALE 1:10
door,
fibre cement on MDF
tap
glass, green
glass, green behind
waste water channel
spout
MDF support board
fibre cement sheet

Video bar, Stuttgart, Germany

Design: Bottega + Ehrhardt Architekten, Stuttgart

Anthracite through-coloured bar counter and video gallery

For this project, a video bar, Stuttgart architects Bottega + Ehrhardt were not only designers; they were also project developers and clients. The concept was developed together with other partners, who were restaurateurs, and a PR agent. The name "Suite 212" is taken from the first video work shown on TV of the South Korean video artist Nam June Paik.

The 170m² bar on the ground floor of a 1950s office building impresses because of its stringency of form and flexibility of use. On three sides the room is bordered by large display windows, which the architects incorporated into the architecture of the new video bar. A free-standing steel staircase in the room leads to a somewhat smaller room used as a club on the top floor. The colour concept – dark screed, black ceiling, built-in fittings made from anthracite-coloured fibre cement and seating in muted tones – sets the mood in the evenings, whilst during the day the large windows are particularly eye-catching.

These display windows, a legacy of the previous occupier, the ADAC motorist organisation, became the key element in the design. The architects use them in the same way as an over-sized box window to frame the views from above and below. The existing heaters under the windows are hidden by continuous bench seating. Likewise the technical services are concealed by the ceiling. A series of monitors showing the daily selection of videos is installed above the bar counter.

For these fittings Bottega + Ehrhardt chose 8mm thick, anthracite through-coloured fibre cement sheets, which are glued to a wooden substructure. Particular value was placed on the homogeneity of the surface and on the authenticity of the material. As the colour was also visible on the cut edges the sheets could be simply butted together and yet still retain the appearance of homogenous colour. In contrast, the tangible presence of the furniture, smooth and stone-like, is emphasised.
With one eye on the durability of the material, here enhanced with a water-repellent coating, the architects chose fibre cement for the cladding material for the bar counter as well as for the walls of the ancillary rooms. The anthracite through-coloured fibre

Black steel staircase in fibre cement space

Show window as a seating landscape

Hard shell – soft centre

cement sheets of the long bar counter are elegantly framed with panels of solid mahogany. Behind the bar, aluminium shelves cantilever from the fibre cement wall and are directly attached to the wooden substructure. The fibre cement sheets, which were put in place later, conceal the fastenings so that the shelf supports cannot be seen. On one side the counter departs slightly from the orthogonal where it takes up the direction of the passageway to the ancillary rooms and leaves more space for the DJ. Finally the bar counter disappears into the wall formed in the same material, behind which lie the staff and storage facilities. A full surface mirror at the end of the passageway creates the illusion that it leads back to the bar. Suite 212 is different to many other bars because it began with a well thought-through design concept: from the concept for the rooms through the materials used to the entertainment facilities. The architects succeeded in giving an impetus to the improvement of an unattractive part of downtown Stuttgart. In the meantime the district has gone on to become a well-known bar mile with other trendy locations and the former six-lane main street has been restored.

Photographs: Alexander Fischer, Ostfildern

Material combination: fibre cement and mahogany

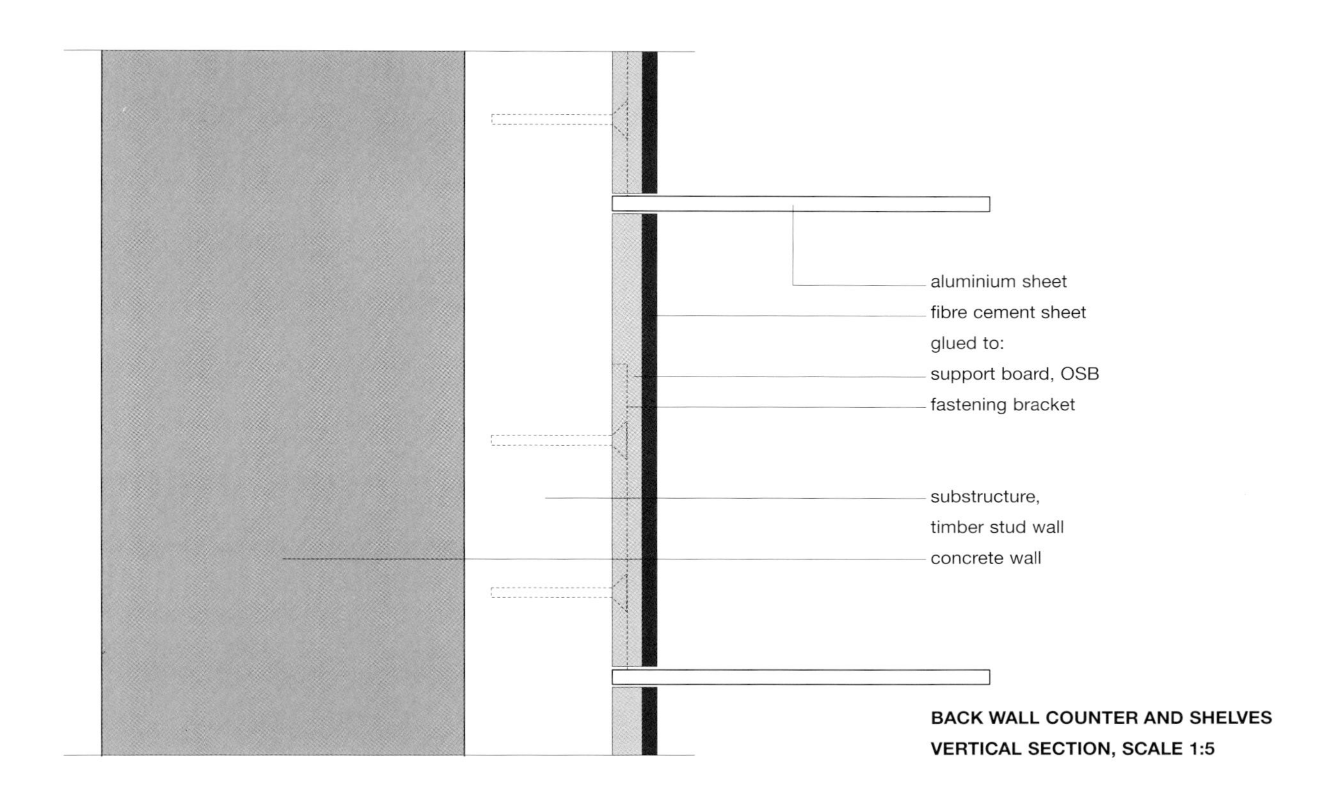

BACK WALL COUNTER AND SHELVES
VERTICAL SECTION, SCALE 1:5

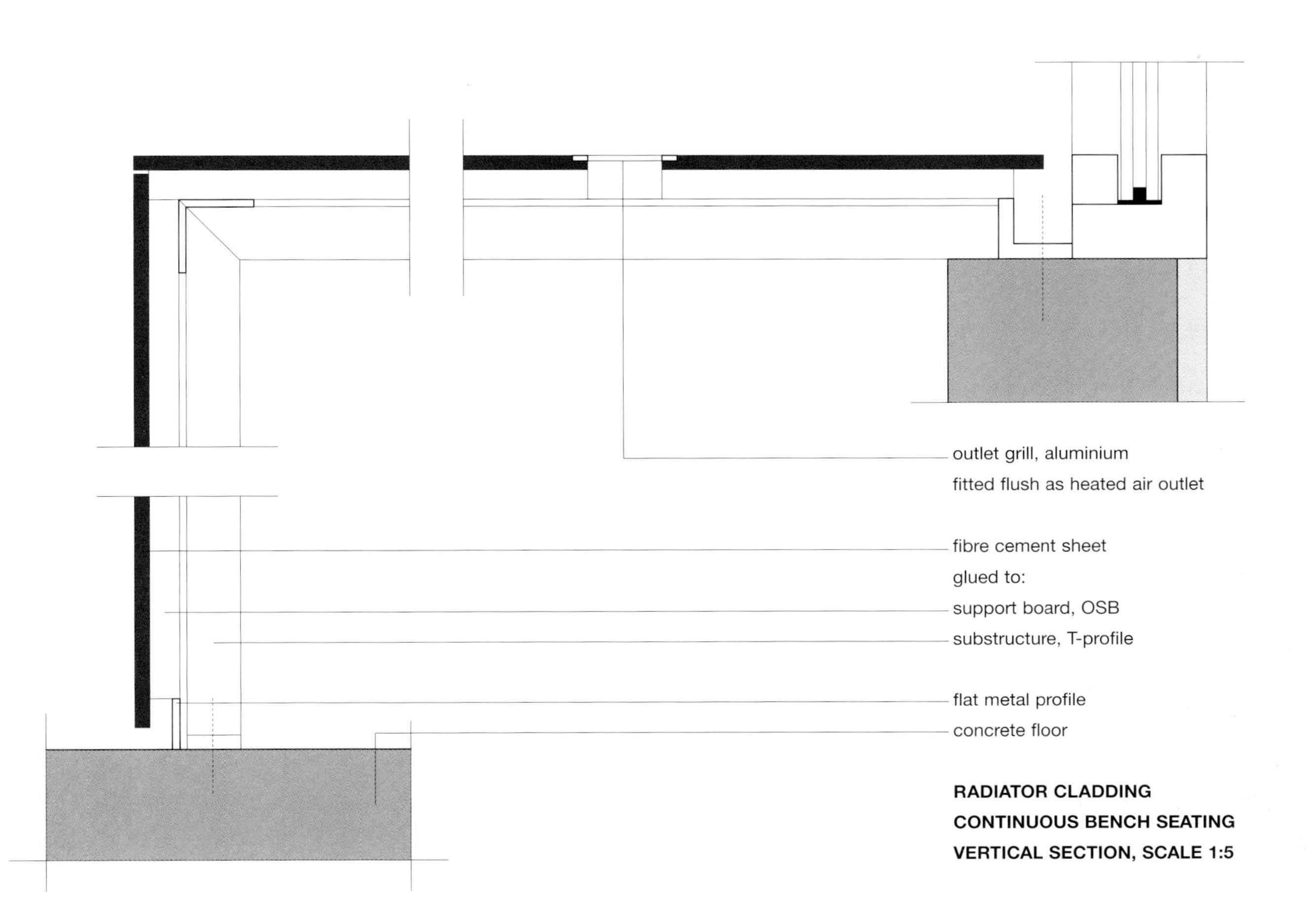

RADIATOR CLADDING
CONTINUOUS BENCH SEATING
VERTICAL SECTION, SCALE 1:5

Cabaret Voltaire cultural centre, Zurich, Switzerland

Design: Rossetti & Wyss Architekten, Zurich

A cultural centre with a new self-confidence has sprung up at the birthplace of Dadaism. Artists, writers and playwrights experienced the freedom of Dada with readings and performances for five months in 1916 at Spiegelgasse 1, Zurich. From the outside the building in the old town scarcely looks any different from its neighbours with their shops and apartments. Inside, however, before reconstruction it had only empty rooms, some of which had been stripped out and their fabric destroyed. Artistic and architectural evidence of the former Cabaret Voltaire was being lost. Therefore the brief for the competition was strict: it called for proposals to record and transform the building and create an experimental, contemporary reference to Dada. The idea submitted by the competition winner Rossetti & Wyss was simple and radical: to extensively expose and preserve the raw fabric of the space. All visible refurbishment would relate solely to the additional fixtures of the new service facilities. For this the architects implemented a consistent change of material and returned to a medium that was as avant-garde in the Dada era as the movement itself. All interiors are constructed using the most advanced form of fibre cement sheets: large format and in various through-colours.

Red box: bar

The architects' intervention concentrated on the connections between the existing rooms. Coloured boxes with new functions were implanted at three nodes: the shop, library and café bar. As organising elements, they form a counterpole to the adjoining rooms left open to transformation. Old masonry courses, decorative residues and art by Dadaist followers remain as visible traces of the previous use and form a contrasting collage. The architects have consciously avoided fashioning the customary perfect finish for this renovation. Some visitors may ask whether the refurbishment has been completed. The journey to the revival and rediscovery of this place starts with this question.

Whoever enters the building from the Münstergasse finds themselves in the front exhibition room, the so-called Showcase. On from that room, an anthracite box with the till counter and shop welcomes the visitor. The route leads through this box like a canal lock down to the vaulted room. In the box, following the counter which has display cabinets for books and smaller items, several steps bridge the difference in levels in the building. All elements of the boxes are constructed from fibre cement: the walls, steps, handrails and counters. The sheets are fitted to one another accurately, glued to a wooden substructure, only at the ceiling are additional concealed screws used. The furnishings combine with the room within a room and become an integral part of the box or transform the box itself into furniture. Directly at the entrance, a flight of stairs leads up into the yellow fibre cement box. The stair goes up through the ceiling and turns into book shelves. On the top floor there is a reading room between the book shelves and the bar in which to sit and relax. Designed as an implant of red, through-coloured fibre cement sheets, the bar provides an appropriate setting for the long traverse to the Cabaret Voltaire and serves both sides equally well.

An effective diversity of light forms plays an integral part in the three fibre cement boxes. Strip lights are inlaid like marquetry from the wall into ceiling and articulate the elongated room. All the steps are set off with spotlights, the fibre cement bar rises on an illuminated plinth, the recesses have atmospheric

Yellow box: stairs and library

Anthracite grey box: shop

Yellow box: stairs

lighting. An unconventional treatment of the surface is used to give a special effect. Grinding produces a soft texture and brings out the unusual grain of the fibre cement. The gentle appearance of the new surface offers a visual and haptic contrast to the rough existing fabric of mixed and faded colours, crumbling plaster and exposed masonry walls.

The implants are service elements in two senses. In addition to their actual function, the view they provide helps to boost the interest of the main rooms. Their pure form means that they do not in any way steal the show from the historic building, instead they create a new awareness of the vivid past of this legendary place.

Photographs: Lorenz Bettler, Zurich

Reading room

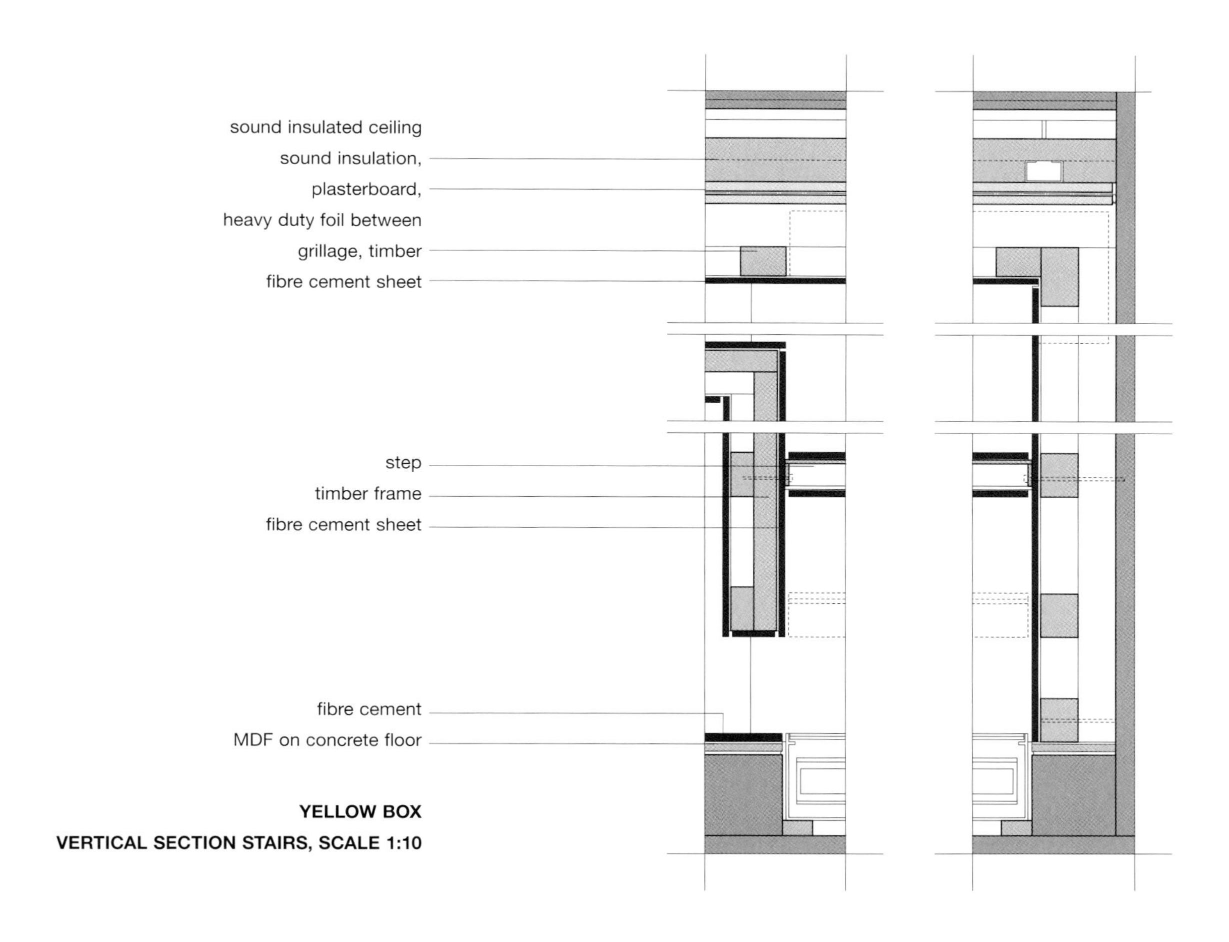

YELLOW BOX
VERTICAL SECTION STAIRS, SCALE 1:10

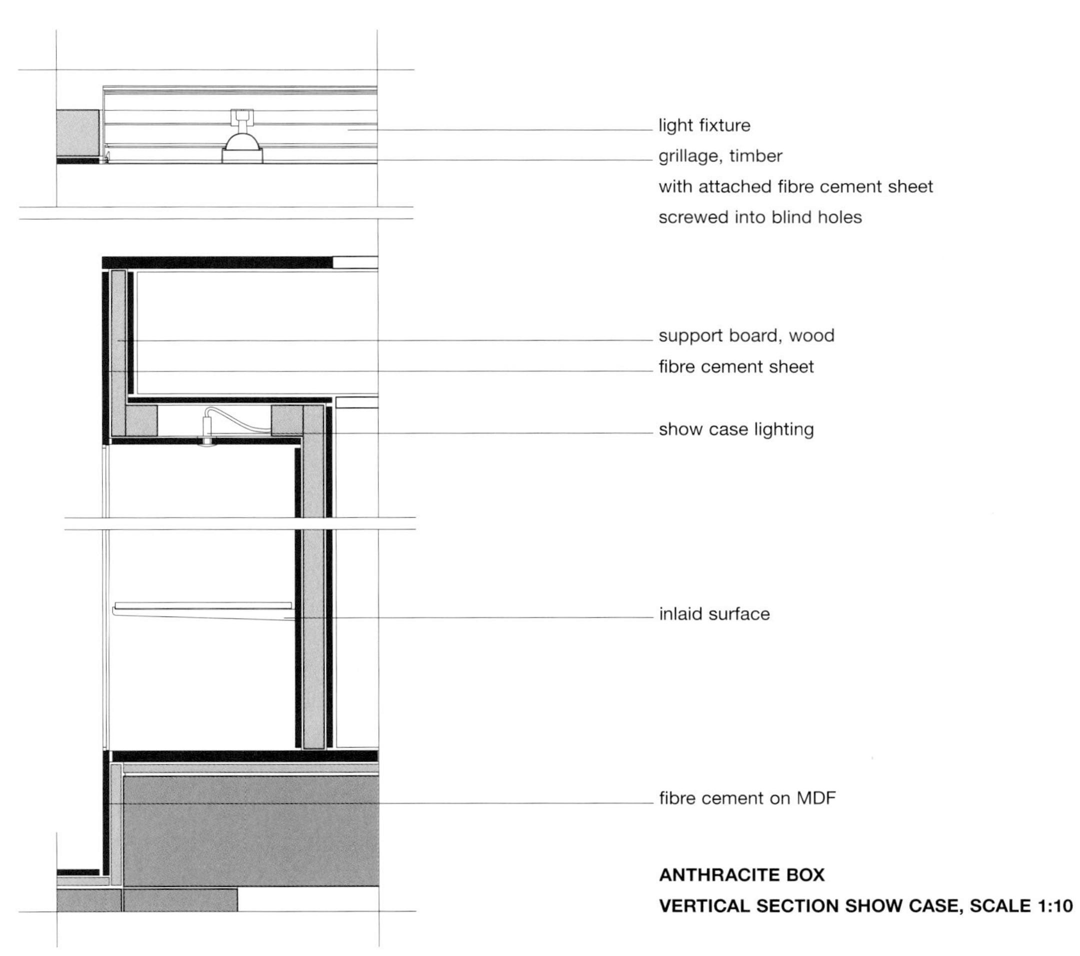

ANTHRACITE BOX
VERTICAL SECTION SHOW CASE, SCALE 1:10

Museum for children in the Elias Church, Berlin, Germany

Design: Klaus Block, Berlin

Colour and material contrast: inside light wood – outside fibre cement

Hardly any other building project demands such a degree of sensitivity and integrity from everyone involved as the conversion of a redundant church for a new, secular use. Berlin architect Klaus Block has completed many excellent church conversions and is a specialist in this type of work. He is also responsible for the transformation of the listed Eliaskirche in the Berlin district of Prenzlauer Berg into a museum for children.

The main attraction of this museum is its three-dimensional climbing labyrinth. The climbing labyrinth is inserted into the middle of the nave and is designed as a self-contained, removable component, which means it could be dismantled and taken away without any problems at the end of the 75 year lease. Free-standing in the room, it consists of two vertical maze walls connected by narrow bridges. The gap between the two walls is on the line between the church entrance and altar. The external linear skin of the labyrinth is made of natural grey fibre cement sheets, which stand in clear, calm contrast to the wooden interior. On entering the church, the first thing the visitor notices on the ground floor is the slender exposed concrete supports and the fibre cement sheet cladding to the underside of the newly introduced floor on which the climbing labyrinth stands.

The labyrinth is entered though the chancel. From here the stairs in the installation go down to the ground floor. Under the stairs, to the left and right of the gap, the red linoleum floors lead in a gentle sweep to the ceiling and open up the way to the apse. The characteristic wall mosaic of the chancel is retained. Wide wooden stairs directly lead there from the narrow sides of the labyrinth. The children are meant to feel confident enough to clamber up each tall step and reach the top. Nearby seating steps change the stairs into a spectator stand.

On reaching the top, the clamberers find themselves high up in the former gallery. Only from here does one realise that the labyrinth extends almost up to the ceiling of the high church interior. Whilst the closed exterior surfaces are clad with large format fibre cement sheets attached by screws, the exposed surfaces are clad with narrow fibre cement strips. The open sides

are made secure by fine steel mesh only. The internal construction is a steel and wood frame, fully boarded out with light-coloured pine.
The fibre cement cladding transforms the timber frame into a sophisticated block. The labyrinthine furniture is designed as a house within a house. Whilst the illuminated, yellow, wooden indoor surfaces radiate their atmosphere outwards, the bare, uncoated fibre cement sheets look unpretentious, as in a similar sense the church interior itself appears roughened and used by the ravages of time. It is hardly discernable how much refurbishment effort would be required within these historical walls in order for the interior to look like a sacred space again. With its modern language of form, the climbing labyrinth introduces a fresh accent and allows the qualities and history of the church interior

Room-high climbing labyrinth with fibre cement cladding

Ravine and bridges

Monolithic appearance

to be experienced in a new way. The details of the original church construction can be seen at close quarters from the higher levels, which otherwise would be almost indiscernible from the floor of the nave.
In another respect the sensitive conversion of the church into a museum was a fortunate event, as it had already been proposed to establish a bank branch there. With public support the children and youth museum could finance the reconstruction and recreation of a prestigious building on this site.

Photographs: Ulrich Schwarz, Berlin

Room-within-a-room structure

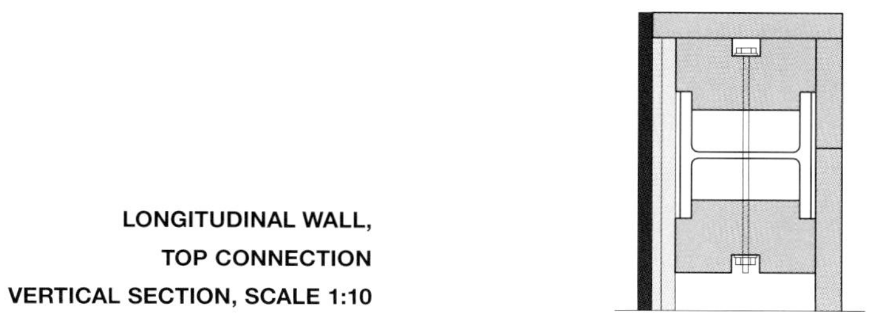

LONGITUDINAL WALL,
TOP CONNECTION
VERTICAL SECTION, SCALE 1:10

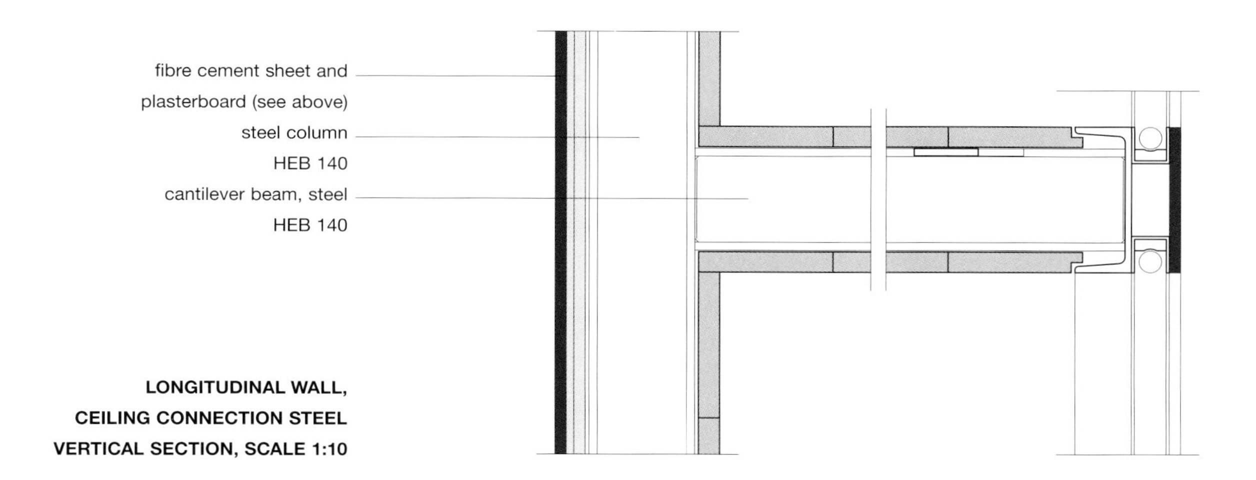

LONGITUDINAL WALL,
CEILING CONNECTION STEEL
VERTICAL SECTION, SCALE 1:10

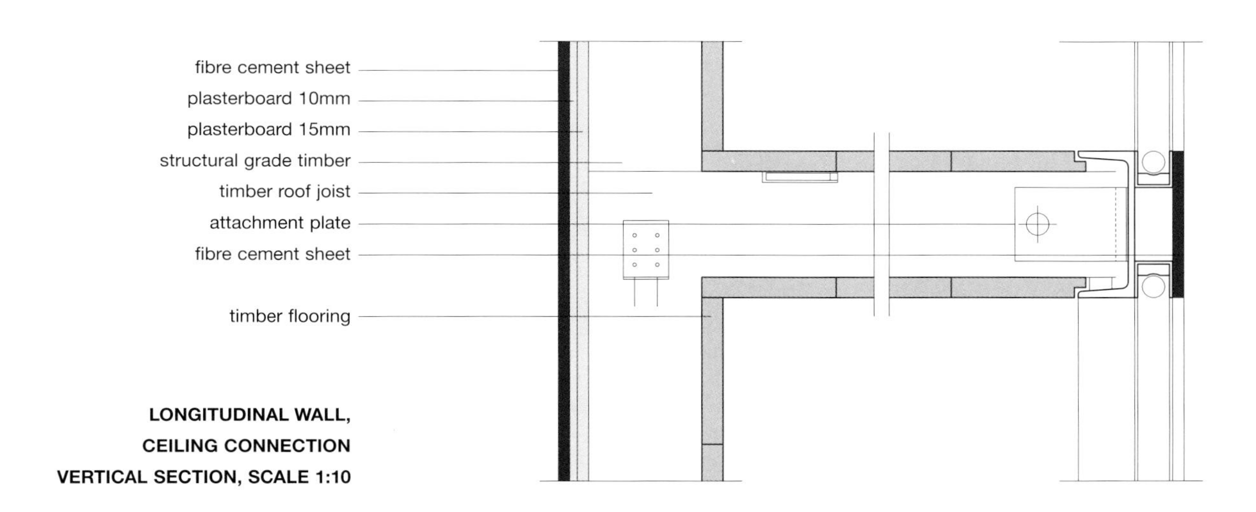

LONGITUDINAL WALL,
CEILING CONNECTION
VERTICAL SECTION, SCALE 1:10

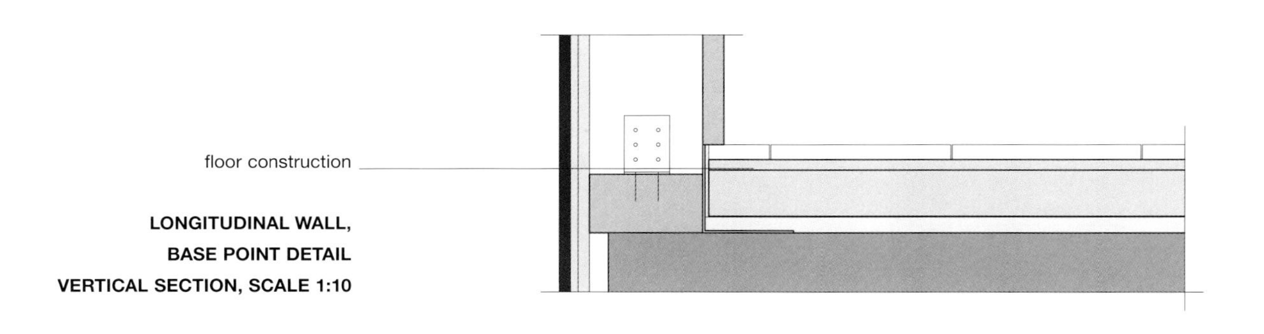

LONGITUDINAL WALL,
BASE POINT DETAIL
VERTICAL SECTION, SCALE 1:10

Staff restaurant, Niederurnen, Switzerland

Design: Cadosch & Zimmermann, Zurich

Kitchen cockpit: free-standing counter of red fibre cement

In retrospect it seems obvious that the refurbishment of the staff restaurant at Eternit AG should be completed using fibre cement sheets. Nevertheless, the architects' choice of the traditional roof and facade material for the interior of a canteen was a courageous decision because the manufacturer had no significant experience of its use in this type of area.
In the 100 year old pavilion construction on the ground floor of a former office building, the lightweight material offers a contrast to the solid fabric of the old building. Even the sculptural roof of the historic pavilion receives a new covering of anthracite-coloured fibre cement roof sheets which is easily adapted to the curved roof shape. Red through-coloured fibre cement slates extend over the full storey height on the entrance side around the building. Like a second skin, the thin, finely-profiled layer closely follows round the corners of the building. The 8mm thick, large format sheets sit on vertical top hat sections and the narrow joints are left open. They connect directly to the door and show window with light metal frames. They have a purely design function in front of the white, rendered,

Fixed point in the room

solid wall of random rubble masonry and refer to a new meaning of the historic building.
The entrance and stairs rely on colour and light. Several of the building's former stairways were consolidated into one main stairwell. One wall in the entrance area is clad over the full room height with blue through-coloured sheets. The upright, anthracite through-coloured sheets at the stairs accentuate the vertical orientation of the space. On the inside of the stairs the side walls are painted a warm shade of red-orange. The brushed chromium steel stair handrail has integral indirect lighting units, which brighten the colour and illuminate an inviting way up the stairs, which are covered with a liquid-applied plastic coating. The grey shades of the fibre cement and the exposed concrete soffits of the stairs combine coherently.
Visible through a glass door, the main self-service fittings and equipment announce the dining hall. The angled shape of the free-standing counter separates the open kitchen spatially from the dining hall and connects both in a single function and communication space. The counter and the suspended superstructure over it are both finished in shiny chromium nickel steel and clad with the same ruby-red fibre cement sheets as the facade at the entrance. The large 104 x 224cm sheet format selected allows the number of horizontal and vertical joints to be greatly reduced. Some of these joints are taken up and continued by the chromium-plated service elements. The pronounced linearity emphasises the longitudinal

Walls clad with anthracite through-coloured fibre cement sheets

orientation of the catering furniture in the room. The fluorescent lighting tubes integrated into the upper part of the counter illuminate the intermediate space up to the ceiling. The architects also use colour here: coming from the kitchen, the walls are covered with a blue textile, on the other side with yellow.
The work of the architects excels in its strong, linear tectonics and careful detail in dialogue with the existing building. The through-coloured fibre cement moves the material into the foreground without overplaying the product theme. The red, blue and anthracite through-coloured fibre cement sheets are neither decoration nor a product show from the manufacturer, but rather fulfil real, space-defining and orientational functions. Here, inside the building and in close proximity, the deep brilliance of the colours and the special quality of the surface are perceived much more intensively than at the facade.

Photographs: Jürg Zimmermann, Zurich

Material combination: fibre cement and stainless steel

Red sheets display corporate identity and refer to the new interior

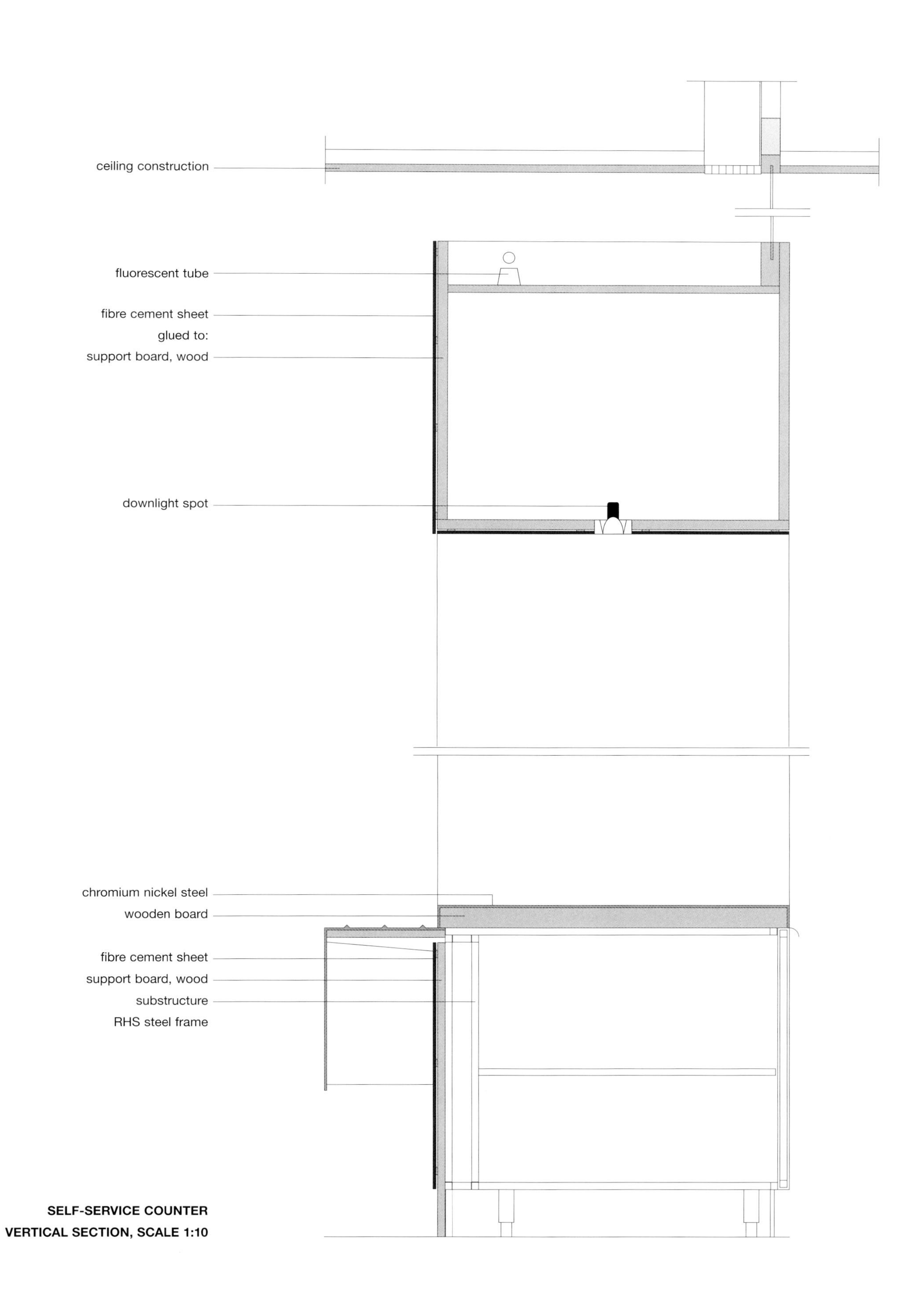

SELF-SERVICE COUNTER
VERTICAL SECTION, SCALE 1:10

Design with fibre cement

Design is not just a will to form. Design also involves the love of experimentation with a material, its properties, possibilities and its inherent strength. Of particular interest are industrially processed materials with the potential for series production. Design is a balancing act between the fantasy of the designer and the technical constraints on the material. Not infrequently the designer succeeds in this context in making the most of what is possible within these constraints, overcoming the boundaries and achieving undreamt-of results. The material has provided a theme for design professionals since well before Swiss designer Willy Guhl formed his 1954 fibre cement beach chair, the legendary and classic “Loop”. Le Corbusier had designed fibre cement furniture 40 years beforehand. The velvetiness of this stone-like material and its free formability make it attractive to today’s architects and designers, who are using the cementitious material to create furniture and objects. Fibre cement’s frost resistance is perfect for plant containers for garden, piazzas and landscaping. There is no shortage of ideas for using fibre cement inside as well: anything from shelves to office lighting to wash basins. Sometimes as prototypes, sometimes in series production – always hand-crafted: every piece is unique. As well as showing fibre cement semi-finished products such as plain, untreated sheets or plant containers, which are promoted to design objects in this new context, the selected examples also include three-dimensional shapes created with thin, formable fibre cement layers.

Archive cabinet

Design: hpk + p, Düsseldorf

Ever since its introduction fibre cement has been used for individual furniture designs. As early as 1915 Le Corbusier designed a cupboard out of fibre cement, wood and linoleum. A contemporary example of fibre cement furniture is the video achive wall, "Locomotion", designed by Düsseldorf architects hpk+p/Heider Pannen-Vulpi von Kalben. The client, a company engaged in the post-production and editing of advertising films, was looking for a special solution for storing its video archive. The architects developed the concept of designing the archive like a wall. They combined the visual character of an exposed concrete wall with the functionality of wall cupboards. What looks like solid exposed concrete when closed proves to be

Lively surface texture of oiled fibre cement slates

Visual character of an exposed concrete wall

only a thin, 6mm thick, and correspondingly light front of fibre cement sheets.
The room-high cupboard occupies the whole wall, with the result that the area and depth of the archive are scarcely perceptible. There is space for 3500 video cassettes. The archive was designed using the same principle as an apothecary's drawer to make best use of the 70cm cupboard depth. The user operates the drawers from the side and pulls them out individually. The system accommodates all formats of video, whilst saving space and ensuring that they can be easily found. A regular 30 x 30cm grid divides the 5.60m wide by 2.70m high front into 98 drawers, which open with push-button switches.
The drawer fronts were cut to shape in the factory and glued to an MDF backing. The architects deliberately avoided handles and labels so as not to disturb the visual effect of a homogenous wall. The surface treatment of the material is of particular interest: a coated finish would have been too uniform for the architects. Instead they oiled the untreated surface of fibre cement panels to prevent signs of everyday use, such as fingerprints, from appearing. The oil penetrates to different depths into the fibre cement sheets and creates a lively surface texture.
The client also uses the video archive room to show videos to customers. The wall opposite the cupboard is used as a video screen. Thick curtains draw to darken the room during the day. The rest of the surfaces in the room are given warm colours to reduce reflections. The cement-grey archive cupboard forms an exciting contrast to the fine surface of the red-brown wooden floor. The prestigious atmosphere of the building is maintained by this restrained design.

Photographs: hpk + p, A. Oemichen, Düsseldorf

Pendant luminaire

Design: Rupert Kopp | Greige, Berlin

Suspended luminaire hand-formed from fibre cement

With his design of the pendant luminaire "Evio", product designer Rupert Kopp departed from the well-trodden path in the use of materials in a complex, year-long development process in dialogue with a lighting manufacturer and a fibre cement manufacturer – two industries that would never meet one another under normal circumstances.
The designer developed production-ready elements with both partners for a pendant luminaire of unexpected elegance and precision. The fibre cement housings flow around each side of the technical corpus of the luminaire to form a lightly curved and dynamic-looking shape. The slender form was made possible by a particularly compact cell louvre, an innovative lighting technology from the lighting manufacturer. With clear lines the housing meets the technical requirements: the corpus shape shows where louvre or starter requires space. The construction and function of the pendant luminaire are evident. The design becomes the principle – even on the often neglected luminaire top.
The unusual use of the traditional roof and facade material for a luminaire housing and the resulting contrast is what makes the special appeal of "Evio". Rupert Kopp's idea was to design a delicate object out of rough construction material, a luminaire that in its materiality borrows from building rather than from furnishing. The housing is available in two materials, one version in fibre cement, the other in plywood – two materials which, despite their differences, can be made into similar flowing forms and fit so well into architecture. What both materials have in common is a liveliness of surface texture, grain and colour nuances.
Like all other formed fibre cement items, this luminaire is manufactured by hand. Each copy is unique. To achieve the desired precision the designer drew upon his experience of working with moulded wood-plastic composites. After various trials with thicker and therefore stiffer materials, the conclusion was reached that the shell should be as flexible as possible and hence should be kept as thin as possible. Louvre and shell support one another: an aluminium strip on the fibre cement shell engages with a similar

Sensual shape and haptic

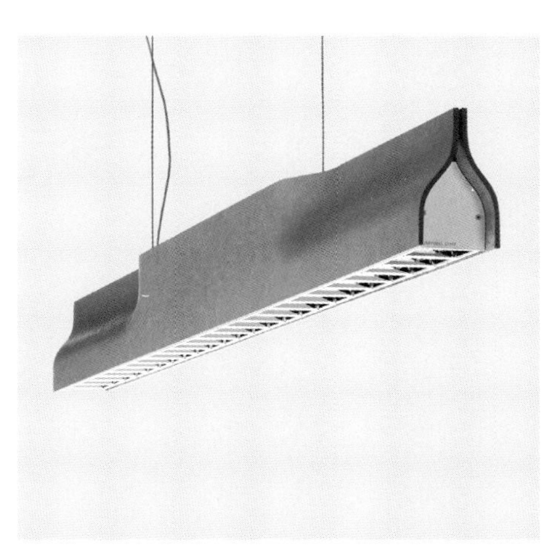

Direct light downwards

strip on the louvre. The flexible shell, made of only 4.5 to 5mm thick fibre cement, is stiffened by the aluminium profile so effectively that its bottom edge remains straight.
Visible connections where materials join would have destroyed the elegance of the form. Therefore the shells were innovatively glued with a special adhesive to a central profile. The fibre cement remains safe and dimensionally stable even at the high temperatures of up to 70°C reached by a luminaire.
There are two sizes of "Evio" available: a slim version for use as a display light for direct illumination downward. In the somewhat longer and wider version the two shells are moved apart slightly. The 20mm wide light-emitting area on the top of the luminaire is used as a light transmissive diffuser. The indirect light directed upwards provides even ceiling illumination at the workplace. The innovative spirit of the design is best shown by the effect the luminaires have on the space. Their stone materiality means they are less of a technical accessory and more of an integral component of the architecture.

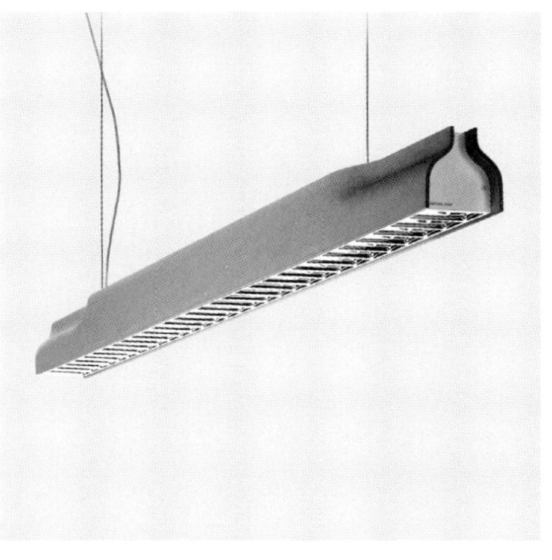

Indirect light upwards

Photographs: Zumtobel Licht GmbH, Lemgo

Shelving and sideboard

Design: Michel Bruggmann, Münchenstein

Original idea: Balcony plant container cut to form sideboard

Whilst looking for a bookcase in a DIY store, Swiss industrial designer Michel Bruggmann discovered some plant containers made from fibre cement. That gave him an idea: with the bottoms removed the plant containers would make good furniture. He approached the manufacturer, sketched a design from which a first prototype was built and was subsequently developed into a range of more than ten products, including lights, speaker cabinets and office furniture. His bookshelf range "Booky", for example, has since sold thousands in Switzerland and has been shown at the Milan Furniture Fair.

In the meantime, the "Bookies" are no longer manufactured from mass-produced plant containers but have become a hand-finished, series-produced product in their own right. Michel Bruggmann perfected the process. The dimensional tolerances of the simple garden tubs did not match those demanded for furniture design. Plant tubs are cast in an external mould, and the resulting dimensional variations on the inside surfaces made the precise fitting of shelves and doors difficult to achieve. Therefore Bruggmann developed a process in which fibre cement was moulded from the outside around a model. The young

Storage cupboard with balcony plant container base as a door

designer succeeded in convincing the manufacturer's development department of the feasibility of his idea. The idea that the designer came up with to support the shelves and guide the sliding doors was as simple as it was elegant: the boxes would be lined with ribbed carpet.

Well before his first encounter with fibre cement, the designer had shown an affinity for freely formable materials; his previous experiments had included natural fibres bonded with resin and starch. The portfolio of fibre cement products, which he developed in conjunction with architect Stephan Eicher, ranges from the "Stella" floor lamp to the "Sitty" seat.

Another particularly individual interpretation of this material is his "Findling": a stone seat made from fibre cement. The shape of this artificial rock was developed by the designer using photomontages on his computer. It has natural-looking humps and hollows as if the stone had been shaped through water and ice. However, they had been precisely formed to offer comfortable seating in a variety of positions. After two test versions, which entered small-scale production, the shape was further refined and eventually ended up as a 35kg fibre cement rock now found in many gardens. The "Findlings" are shaped by hand – this time traditionally with an external mould and a hollow core – so that in spite of having the same basic shape they are all slightly different.

Shelving made of stacked balcony plant containers without bases

Photographs: Friederike Baetcke, Basel

Wash basin

Design: Astrid Bornheim, Berlin

Sensual experience of fibre cement

A wash basin made from fibre cement takes the material to the limits of its formability and material suitability. That was what attracted the Berliner architect Astrid Bornheim. Form-finding and product research for this application were carried out as a materials research project and as a further development of the design principles of the showroom and training centre for Eternit in Heidelberg (see page 120), with the aim of making the material capable of being experienced sensually in terms of its spatiality and universality. How can the commonly accepted limits of the material and manufacturing process be expanded? And how can modern computer-aided form-finding processes be combined with traditional, analogue manufacturing techniques? Craftsmen at the Eternit factory worked closely on these issues with architecture students from the Technical University of Brunswick (TU Braunschweig), where Astrid Bornheim teaches experimental design. The forms were developed by modelling and were validated with computer shadow studies. The final form takes into account functional considerations, such as the directional flow

Fibre cement landscape of light and shadow

of the water and the specific requirements of the manufacturing process. Checking the radii of curvature was critical to ensure the curves flowed smoothly from one to the other. The CAD drawings were transformed into a wooden form from which the GRP mould was made. This GRP mould allowed the basin form to be repeated and ensured that it could be reproduced in series.
Modelling the basin in wet fibre cement mesh required much experience, to form the object with an even thickness, without compression or excessive deformations occurring. After one day the cement has cured enough for the basin to be taken out of the mould. After three further weeks the material has hardened sufficiently through its depth for the coating to be applied.
In several series of tests a special coating process was developed at the Technical University of Brunswick that allows the character of the material to show through and retains its naturalness. Several layers of acrylic lacquer are applied after any cement film from the production process has been ground smooth. The intermediate coatings are ground to smooth any fibres that may have become raised. A final wax polish gives the surface a matt, silky feel and makes the wash basin more water-repellent. The polish must be reapplied at regular intervals.
The result is a functional article and a sensual landscape of light and shadow.

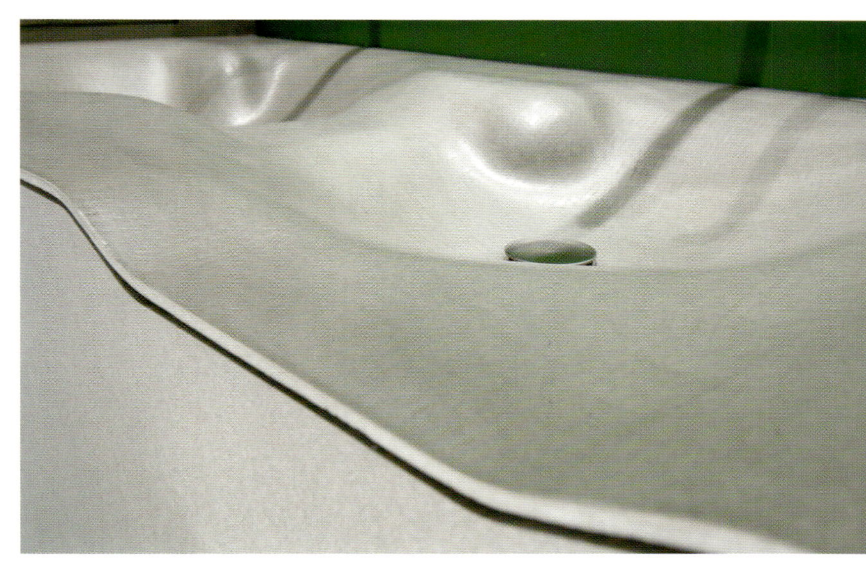

Curves flowing smoothly into one another

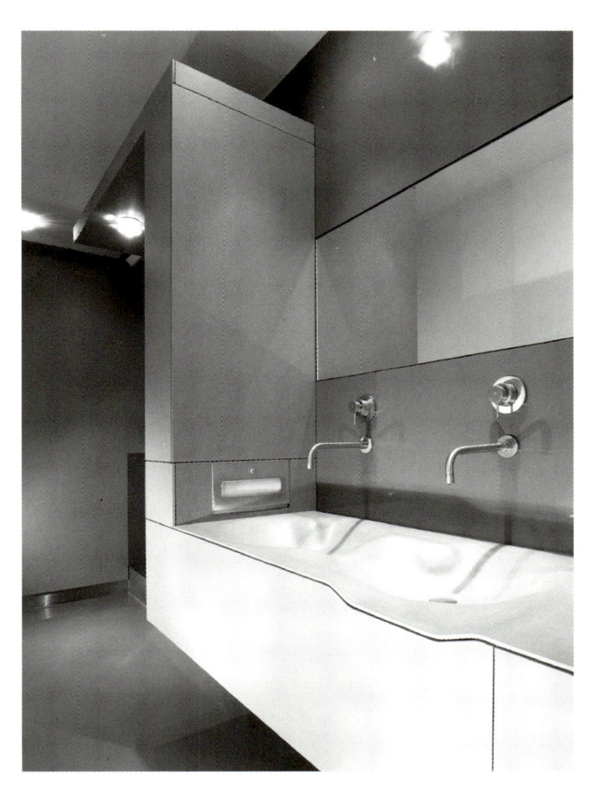

Composition of flat slates and free forms

Photographs: David Franck, Stuttgart

Garden courtyard, University of Stuttgart, Germany

Design: Pfrommer + Roeder, Stuttgart

Fibre cement chair on a tile mosaic in the "Persian courtyard"

The fibre cement beach chair from Swiss product designer Willy Guhl counts among the classics in modern furniture design. Even after 50 years it has lost none of its contemporary appeal, as is demonstrated by the Stuttgart landscape architects Pfrommer + Roeder in their design for the new building at the Institute for Computer Science, University of Stuttgart: they treat the fibre cement chair like an elegant piece of furniture and place it on a "Persian carpet" of tile mosaics. Small occasional tables from the same range of furniture, three-armed lamps and giant flower pots further into the courtyard complete the picture of an abstract outdoor living room.

The "Persian courtyard" is one of four courtyards in the new institute building. Following on from the classic element of the living room, all four courtyards were designed to have their own individual character. The common theme of the living room is intended as a contrast to the highly technical world of work at the Institute for Computer Science. The landscape architects installed lounge-like outside spaces and used simple and effective means to represent four types of carpet – Berber, Persian, sisal and flokati – on accessible platforms. The furnishings are in keeping with this style and by experimenting with the proportions of the carpets, chairs, lamps, potted plants and people, the landscape architects create a bizarre atmosphere.

For the "Persian carpet" the landscape architects investigated the classic motifs of this traditional knotting art and finally selected a motif, which they reduced to eight colours on the computer. The resulting pattern was used for a floor mosaic made from square tiles. By using glazed and unglazed mosaic tiles they created a floor surface that has a special shimmer in sunlight.

As a modern interpretation of the traditional motif, the landscape architects dispensed with the ornaments and symbols that would normally be the formal highlight at the centre of a real Persian carpet. The stylised garden motifs and ornaments of Persian carpets often make reference to paradise or the Garden of Eden. The edges of the carpets symbolise boundary walls and paths. Pfrommer + Roeder took up this

Chairs and tables arranged in a row or individually

idea and put it into words in an informally positioned script: "pairi-daeza" (the enclosed garden) can be read on the matt black surface. The script in mosaic blocks of the same colour shows up only faintly through the glazed surface.

Like a living room the designers have furnished the four courtyards with appropriate, weather-resistant seating elements. For the Persian courtyard they chose the garden chair by Willy Guhl as the embodiment of an elegant white chair. Lightness, stability and sculptural quality are the most impressive features of this fibre cement furniture classic. "I have known the 'Loop' beach chair by Willy Guhl since I was a child and was very happy to discover that there are now small occasional tables as well as seats," explains project manager Hendrik Scholz.

The landscape architects intended to create an austere arrangement of the tables and chairs, reminiscent of a sanatorium or ship's deck. However, this fibre cement furniture is very easy to move, with the result that the users continuously rearrange it, forming different constellations.

For the project "4 Courtyards. 4 Lounge Types. 4 Carpet Themes." Pfrommer + Roeder received a commendation at the German Landscape Architecture Awards. The jury praised the idea: "The joke in the background in the approach to the selected theme of the living room is even more appropriate when one considers the young age and the needs of the users. In this case the users – computer science students –, a target group which as a result of their chosen career will spend a lot of time in front of a computer, are offered these four discrete alternative worlds."

Photographs: Hendrik Scholz, Stuttgart

Furniture set

Beach chair and table

Design: Willy Guhl

Design classic from 1954

Loop formed from a fibre cement sheet

The designer Willy Guhl has had a prominent role in the design history of fibre cement. When he taught interior architecture at the School of Applied Arts in Zurich during the 1950s, he often had objects made of this material designed in his class. A selection of the best designs went into production. Some are still well known today, such as the "Spindle" by Anton Bee. Guhl risked the step from applied arts to industrial design. Not only as a teacher but also as a designer, he thoroughly explored the possibilities of the material, which had until then been used only as a building material and for the manufacture of simple window boxes. His own works, among them the "Loop" beach chair, have contributed considerably to the fame and popularity of the material and moved it into the world of design. As an early proponent of modern

industrial design, Guhl succeeded in designing good everyday objects for production using modern techniques. He analysed the production process and researched the possibilities and limits of the material.
Even before he became familiar with fibre cement, Willy Guhl had experimented with the development of moulded seats made out of a new type of plastic. His aim was to develop a new ergonomic seat shape. He formed 1:1 scale clay models and cast the best of them in gypsum plaster. In the development of his 1954 beach chair he followed the same sequence. The model he used to present his design to the manufacturer was made in a similar way: a lump of solid plaster, which had been appropriately painted by Guhl to suggest the material. Guhl developed the corresponding pattern and created one of the most successful design objects made of fibre cement.
In his form-finding Willy Guhl was inspired by the manufacturing process for fibre cement panels: the curved chair looks like a never-ending belt rolling over an arrangement of rollers, or a freely formed loop from a standard, rectangular plate. The high break strength and tensile strength of the material mean there is no need for any other support.
After the composition of the material was changed from mineral asbestos to synthetic organic fibres made from polyvinyl alcohol, the chair could no longer be manufactured in its original form. The changed material properties meant a redesign was required. On the occasion of his 80th birthday Willy Guhl presented the manufacturer with a redesign of the successful garden chair. The seat and back of the chair needed to be stiffened for manufacture with the new asbestos-free material. The basic idea of a continuous band was retained. This also characterised the associated table that went into production soon after. Like its predecessor, the 1954 beach chair, the weather resistant garden furniture is, appropriately for the material, beautifully shaped, and a perfect example of the versatility of fibre cement.

Occasional table in the same style

Photographs: Eternit AG, Niederurnen

"Shape Memory" exhibition, Berlin, Germany

Design: Heimbach, Bornheim, Fingerle & Woeste

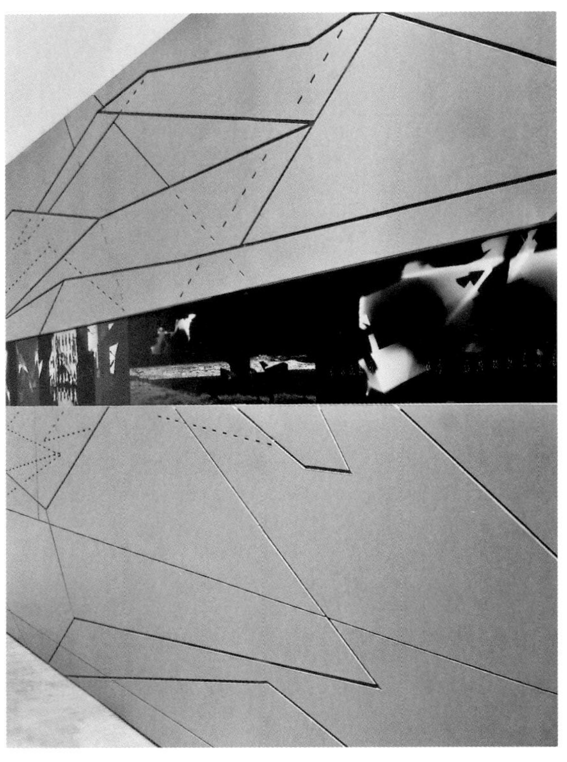

Pattern cut with a water jet

The "Shape Memory" exhibition is an installation project made out of fibre cement slates, which is taking place in several parts and concerns itself with living on an abstract, theoretical level. The title refers to a term in materials science dealing with smart materials and denotes the reversible change in shape of a material by means of a phase change. Also, it opens a field of association between the terms "shape" and "memory". In their room installation, the architects Heimbach, Bornheim, Fingerle & Woeste explore the theme of living as a waiting state. They research living space as a threshold between city and bed. They explore the idea of the house as a machine for slowing down and accelerating in contrast to the idea of the house as an autonomous container. The starting point for their design experiments is a plan of possibilities, an ideal map. This map is blank, like the ideal map used by the snark hunters in Lewis Carroll's poem, "The Hunting of the Snark". Based on this idea the architects mapped out the fundamental components of living: memory and anticipation of actions, energy statuses and types of storage. The models and draw-

Exhibition architecture in fibre cement

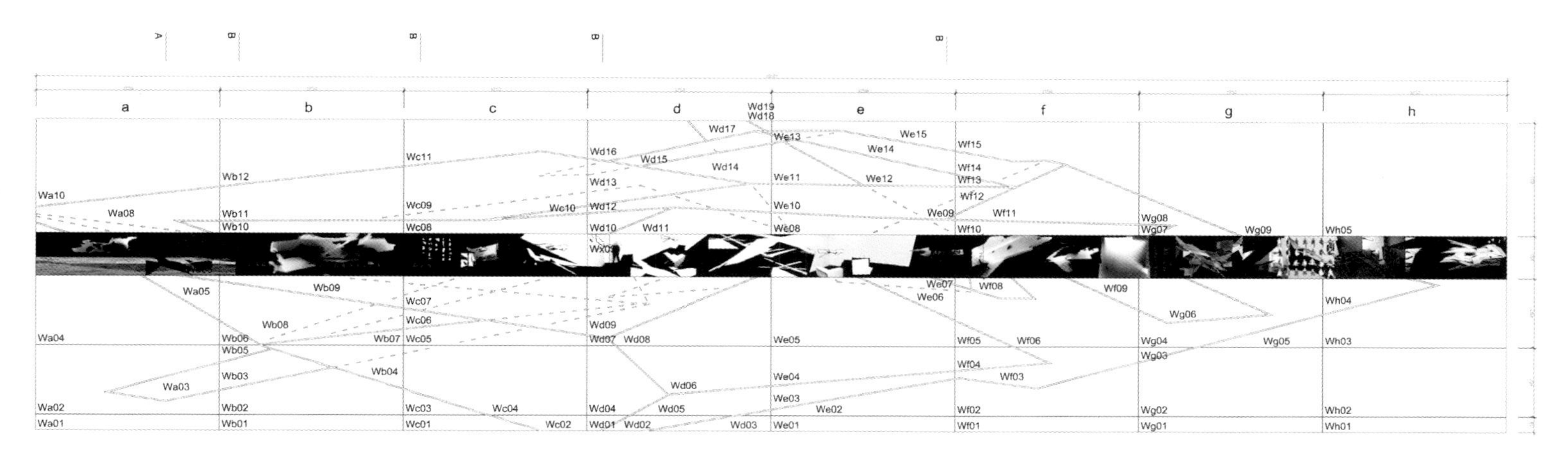

Maps

ings developed from the maps form the bases for spatial installations of wood-cement boards, which are then installed into existing buildings.

In a Cologne apartment in an old building and in a 1950s Hamburg apartment, the architects used their temporary intervention to examine and reinterpret common spatial arrangements and spatial meanings. The project was presented as a panel exhibit made from fibre cement as part of the large Berlin exhibition "Designmai".

A series of images of drawings, photographs and diagrams of the original room installations is built like a horizontal film strip into a 12.00 x 3.10m display of natural grey fibre cement sheets. The sheets form the large two-dimensional passe-partout for the black and white graphics of the film strip. Superimposed on the film strip is a wire-frame drawing, which was cut into the fibre cement slates by a CNC-controlled water-jet cutter. Using this method the architects explored the possibilities of the material and went to the limits of the cutting technique. The individual components are glued over their full surface on to an MDF backing board. The contradiction of the real hardness of the material and its soft visual surface texture made the use of fibre cement seem particularly suitable for this project. The random particle orientation and fine structure of the material invites the observers to immerse themselves in the film strip. The spatial depth of the drawing further emphasises these material characteristics.

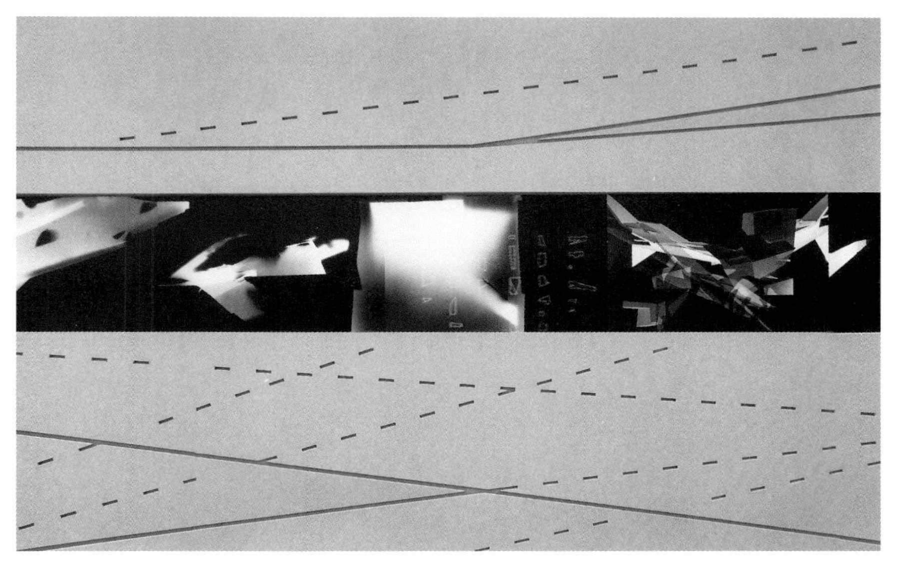

Passe-partout of fibre cement

As could be expected from the title "Shape Memory", each time the installation moves venue, the elements, ideas and memories from the previous exhibition are taken to the next but always in a modified form. At each new venue, the object is made up again and opens a dialogue in scale, form and material with the fresh spirit of its new surroundings. Therefore the fibre cement material selected for the Berlin exhibition was both a reminder and a reinterpretation of the original installations that used wood-cement boards as space-defining elements.

Photographs: Kay Fingerle, Berlin

About the author
Jan R. Krause is Professor of Architecture and Media Management at the University of Applied Sciences, Bochum (FH Bochum). After studying architecture at the Technical University of Brunswick (TU Braunschweig), the Swiss Federal Institute of Technology in Zurich (ETH Zürich) and the Technical University of Vienna (TU Wien) he specialised in professional architecture communication at the College of Journalism, Baden-Württemberg and as Editor and Managing Editor of the architecture magazine AIT. Jan R. Krause lives and works in Berlin, where he heads the corporate communications department at Eternit AG. He writes as a freelance author about architecture and the work of the professional architect for professional journals, daily and commercial press. He has been the chairman of the Deutscher Werkbund, Berlin since 2006.

About the technical drawings
All technical drawings are schematically detailed representations and are intended to illustrate basic construction principles. The construction details may vary according to different requirements and regulations, depending on the country. In every case the applicable standards should be observed and the manufacturer consulted for advice on detailing. All the technical drawings in the chapter on Technology with the kind permission of Eternit AG, Heidelberg. All the technical drawings of the projects with the kind permission of the architects.

About the term Eternit
Eternit has become a synonym and generic term for fibre cement. The manufacturers and distributors of fibre cement products do not always carry in all countries the name of Eternit, other company names include Bachl, Edilit, Eurocem, Euronit, James Hardie, Ivarsson, Landini, Marley, SVK, Tegral and Uralita.

Illustration credits

Hoehn/Laif, Agentur für Photos und Reportagen, Cologne, p. 10
Gerald Zugmann, Vienna, p. 13
Sebastian Hoyer, Dublin, p. 15
Margherita Spiluttini, Vienna, p. 43, 46
delugan meissl associated architects, Vienna, p. 42, 44, 45
Christian Richters, Münster, p. 48-56
Roland Halbe, Stuttgart, p. 58-60
Miran Kambic, Radovljica, p. 62-65
Christian Richters, Münster, p. 66-70
Oliver Heissner, Hamburg, p. 72-74
Michael Weschler, New York City, p. 76-78
Torben Petersen, Denmark, p. 80-82
Philippe Ruault, Nantes, p. 84-86
Ralph Feiner, Malans, p. 90-92
Peter Frese, Wuppertal, p. 94-96
Lukas Roth, Cologne, p. 98-100
Rainer Lautwein, Bochum, p. 102-104
John Donat, Hong Kong, p. 106-108
Jörg Hempel, Aachen, p. 110-112
Alan Jones Architects, Belfast, p. 114-116
David Franck, Stuttgart, p. 120-124
Alexander Fischer, Ostfildern, p. 126-128
Lorenz Bettler, Zurich, p. 130-132
Ulrich Schwarz, Berlin, p. 134-136
Jürg Zimmermann, Zurich, p. 138-140
hpk + p, A. Oemichen, Düsseldorf, p. 144-145
Zumtobel Licht GmbH, Lemgo, p. 146-147
Michel Bruggmann, Münchenstein, p. 148
Friederike Baetcke, Basel, p. 149
David Franck, Stuttgart, p. 150-151
Hendrik Scholz, Stuttgart, p. 152-153
Eternit AG, Niederurnen, p. 154-155
Kay Fingerle, Berlin, p. 156-157

Quotations

[1] Dietmar Steiner: “Architektur Beispiele Eternit – Kulturgeschichte eines Baustoffes”, Löcker Verlag, Vienna, 1994, p. 87

[2] Konrad Wohlhage: Keynote speech at the 75th Jubilee of Eternit Aktiengesellschaft in Heidelberg, 8 October 2004

[3] Werner Oechslin: “Eternit Schweiz – Architektur und Firmenkultur seit 1913”, gta Verlag, ETH Zurich, 2003, p. 7

[4] Adolf Behne: “Der moderne Zweckbau”, Drei Masken Verlag, Munich, 1923

[5] Hannes Mayer: “Die neue Welt”, in: “Das Werk”, No. 7, 1926, p. 222

[6] Max Frisch: “Montauk”, Suhrkamp, Frankfurt, 1975

[7] Wolfgang Pehnt in: “Egon Eiermann – Die Kontinuität der Moderne”, Hatje Cantz Verlag, Ostfildern-Ruit, 2004, p. 20

[8] Fritz Hatschek: “Architektur Beispiele Eternit – Kulturgeschichte eines Baustoffes”, Löcker Verlag, Vienna, 1994, p. 13

[9] Sigfried Giedion: “Die Erfahrungen eines Historikers beim Hausbau”, in: “ac Revue”, No. 6, 1957, Editions Girsberger, Zurich, p. 6